Lakes and Watersheds in the Sierra Nevada of California

Lakes and Watersheds in the Sierra Nevada of California

RESPONSES TO ENVIRONMENTAL CHANGE

John M. Melack
Steven Sadro
James O. Sickman
Jeff Dozier

UNIVERSITY OF CALIFORNIA PRESS

University of California Press, one of the most distinguished university presses in the United States, enriches lives around the world by advancing scholarship in the humanities, social sciences, and natural sciences. Its activities are supported by the UC Press Foundation and by philanthropic contributions from individuals and institutions. For more information, visit www.ucpress.edu.

University of California Press
Oakland, California

Library of Congress Cataloging-in-Publication Data

Names: Melack, J. M., author. | Sadro, Steven, author. | Sickman, James O., 1960– author. | Dozier, Jeff, author.
Title: Lakes and watersheds in the Sierra Nevada of California : responses to environmental change / John M. Melack, Steven Sadro, James O. Sickman, Jeff Dozier.
Description: Oakland, California : University of California Press, [2021] | Includes bibliographical references and index. | Contents: Introduction to the Sierra Nevada — Introduction to high-elevation lakes and watersheds of the Sierra Nevada — Snow hydrology — Watershed hydrology — Watershed biogeochemistry — Limnology — Variability, trends and future scenarios.
Identifiers: LCCN 2020004858 (print) | LCCN 2020004859 (ebook) | ISBN 9780520278790 (cloth) | ISBN 9780520967342 (ebook)
Subjects: LCSH: Lakes—Environmental aspects—Sierra Nevada (Calif. and Nev.) | Watersheds—Environmental aspects—Sierra Nevada (Calif. and Nev.) | Sierra Nevada (Calif. and Nev.) | Emerald Lake (Sequoia National Park, Calif.)
Classification: LCC GB1619 .M45 2021 (print) | LCC GB1619 (ebook) | DDC 551.48/2097944—dc23
LC record available at https://lccn.loc.gov/2020004858
LC ebook record available at https://lccn.loc.gov/2020004859

Manufactured in the United States of America

30 29 28 27 26 25 24 23 22 21
10 9 8 7 6 5 4 3 2 1

*The publisher and the University of California Press
Foundation gratefully acknowledge the generous contribution
to this book provided by the International Community
Foundation/JiJi Foundation Fund.*

*The publisher and the University of California Press
Foundation also gratefully acknowledge the generous support
of the Ralph and Shirley Shapiro Endowment Fund in
Environmental Studies.*

CONTENTS

PREFACE

Lakes have long attracted study by ecologists, and insights derived from their investigation have led to substantial advances in all branches of ecology and ecosystem science. Their scientific value stems, in part, from their position in the landscape. Their hydrological connectivity to the watershed and climatic and depositional connectivity to the airshed make them natural integrators of conditions within their catchments and beyond. Mountain ecosystems, in particular, are valuable systems to study because they possess a high degree of spatial heterogeneity superimposed on steep elevation driven gradients in soil development, vegetation cover, and precipitation and temperature patterns. High-elevation lakes are recognized as excellent indicators of regional environmental changes and are increasingly being used by regulatory agencies to establish critical loads for atmospheric pollutants. Moreover, the importance of montane snowpacks to freshwater supplies mandates that particular attention be paid to the biogeochemistry, hydrology, and ecology of these systems.

California's iconic mountain range, the Sierra Nevada, John Muir's "range of light," harbors thousands of lakes. Most are remote, at high elevations, and blanketed under snow and ice for many months each year. Hence their scientific study is challenging. Concern about acidification from atmospheric deposition in the 1980s spurred efforts to measure deposition and understand its influence on the lakes and their catchments. In more recent decades, limnological research has focused on the role of climate warming and related variations in snowfall and snowmelt in driving change in these ecosystems.

Our goal in this book is to synthesize and integrate over three decades of investigation of high-elevation lakes throughout the Sierra Nevada of

California. We focus on Emerald Lake and its watershed (Sequoia National Park) because no other site has the combined long-term hydrological and limnological data for coupled lake-watershed understanding. Studies have documented recovery from mild acidification after reductions in emissions and have highlighted the vulnerability of Sierra lakes to ongoing changes in climate. Rapid warming of air temperatures and changes in precipitation have modified the timing and magnitude of hydrological and chemical fluxes, altering aquatic ecology. Experiments and models have added a mechanistic understanding of the underlying processes and provided predictive tools for managers. Comparative studies of biological and hydrochemical conditions within lakes scattered throughout the mountain range provide a regional context for examination of Sierra-wide conditions and changes.

The book begins with an introduction to the Sierra Nevada as an integrated system with characteristic climatic, geologic, topographic, ecological, and cultural features. We then introduce the lakes and watersheds that are focus of our studies. To place these lakes in context we present an analysis of the geography of water bodies throughout the Sierra based on a compilation of data from hundreds of lakes.

Chapter 3 focuses on snow, the largest input of water to Sierra watersheds with important influences on many ecological aspects of the lakes. The physical properties of snow and their relevance to remote sensing of snow are introduced. The combination of remote sensing analyses and snow survey data provides Sierra-wide descriptions of the snow cover and its interannual variability. Chapter 4 presents measurements of runoff and water balances for watershed, an essential aspect of the biogeochemical and limnological aspects treated in subsequent chapters. The combination of detailed surveys of seasonal snowpacks and monitored stream discharge and meteorological data allows estimation of all the terms in water balances for high-elevation watersheds spanning the range from north to south.

The fifth chapter examines watershed biogeochemistry based on long-term studies and mechanistic, modeling, and experimental results from the Emerald Lake watershed and elsewhere throughout the Sierra Nevada. The atmospheric deposition of ions and particulate matter represents a significant input to Sierra watersheds, which are sensitive to anthropogenic activities. In turn, the weathering of soils and rocks and the biological uptake within catchments alter the atmospheric inputs and determine loading to the lakes. We discuss biogeochemical mechanisms associated with nitrogen, phosphorus, and major solute dynamics and present mass balances of inputs and outputs for selected watersheds.

Chapter 6 examines long-term limnological and ecological data from Emerald Lake complemented by results from repeated surveys of many lakes throughout the Sierra Nevada and from sediment cores that extend the time back to the 1800s. Seasonal changes in temperature, thermal stratification, and hydrological flushing are presented as these processes establish the physical template for the biogeochemical and ecological processes in the lakes, including rates of primary production and ecosystem respiration. Responses to nutrients and episodic acidification are examined based on experimental and observational data. Diatom frustules and combustion by-products preserved in lake sediments permit an extension of our records of changes in the lakes back in time.

Finally, hydrological, limnological, and ecological responses to climate variability and environmental change are considered in chapter 7. Variations in the seasonal snowpack have a large influence on temperatures in the lake and associated biological processes. These variations in snowfall are driven by large-scale atmospheric and oceanic patterns and are responsive to climate warming.

Our studies of the high-elevation Sierra Nevada began in the late 1970s and were stimulated and augmented by design and serendipity through the following decades. We all share a special fondness for mountains and found the Sierra Nevada a magnet for our scientific interests, aesthetic appreciation, and recreational pursuits. Originally from the eastern United States, John Melack hiked the mountains of the Northeast. On his arrival in California, the Sierra lakes offered him and his students opportunities to conduct comparative limnological and watershed studies in the context of concerns about acidic atmospheric deposition, climatic variability, and episodic events. A fourth-generation Californian, Jeff Dozier hiked every summer in the Sierra Nevada starting from childhood and later climbed difficult routes on mountains on three continents. His initial concern with avoiding avalanche danger led him to the study of snow and its runoff, especially its seasonal and topographic variability at scales of whole mountain ranges. Steven Sadro grew up backpacking in the Sierra Nevada. It was a through-hike along the Pacific Crest Trail that inspired his research interest in lakes and streams in a landscape context. His research uses the steep landscape gradients found in the Sierra as a natural laboratory to explore the effects of environmental and anthropogenic processes on aquatic ecosystem function. Growing up, James Sickman dreamed of being an oceanographer but found a rich outlet for this passion in the watersheds and lakes of the Sierra Nevada. Beginning in his early twenties, he has conducted research designed to assess the vulnerability of the region to human activity.

ACKNOWLEDGMENTS

Many dedicated and hardworking individuals contributed their efforts to this project over the years; we thank Kevin Skeen, Pete Kirchner, Josh Ladau, Tom Smith, Mike Law, Gabriel De La Rosa, Ellen George, Suzanne Sippel, Clifford Ochs, and Stephen Kimura for assistance in the field. Laboratory analyses were a major part of this research. During the course of the project, tens of thousands of samples were processed in analytical laboratories at the University of California, Santa Barbara (UCSB), and University of California, Riverside (UCR), and at field laboratories in Sequoia National Park and at the Sierra Nevada Aquatic Research Laboratory. Our thanks go to Frank Setaro for his dedication in performing chemical analyses and managing the quality assurance/quality control (QA/QC) program for years. We thank Delores Lucero for many years of field assistance and chemical and isotope analyses. Dan Dawson helped in the field and organized much of the operations in the eastern Sierra Nevada.

Especially important were the many graduate students and several postdoctoral researchers who contributed both their labor and their insight: Ned Bair, Leon Barmuta, Adam Cohen, David Clow, Mike Colee, Bert Davis, Kelly Elder, Diana Engle, Steve Hamilton, Helen Hardenbergh, Andrea Heard, Pete Homyak, Pam Hopkins, Annelen Kahl, Rick Kattelmann, Kim Kratz, Al Leydecker, Danny Marks, Lisa Meeker, Amy Miller, Craig Nelson, Tom Painter, Karl Rittger, Walter Rosenthal, Adrianne Smits, Chad Soiseth, Mary Thieme, Will Vicars, Mark Williams, and Michael Williams.

Among the colleagues at UCSB and other universities who assisted and advised us are Robert Holmes, Lance Lesack, Thomas Fisher, John Harte, Scott Cooper, Roger Bales, Soroosh Sorooshian, John Dracup, and Sally MacIntyre. Roland Knapp provided valuable material and editorial advice.

The cooperation and support of the National Park Service personnel at Sequoia–Kings Canyon National Park and Yosemite National Park were essential. In particular, we thank David Parsons, Anne Esperanza, David Graber, Tom Stohlgren, Harold Werner, Jack Vance, Danny Boiano, and Steve Thompson as well as the Lodgepole and Pear Lake Rangers for providing support to our projects. Many of the research sites in this report are in areas under the jurisdiction of the U.S. Forest Service. We would like to thank Andrea Holland, Tom Heller, Bill Bramlette, Neil Berg, Gordon Peregun, and Keith Waterfall for their cooperation and assistance. The Sierra Nevada Aquatic Research Laboratory provided laboratory and living space in support of sampling in the eastern Sierra. Mammoth Mountain Ski Area supported our studies of snow on Mammoth Mountain. Vern Clevenger provided the cover photograph.

We thank John Holmes and Kathy Tonnessen of the California Air Resources Board for their guidance and support over the years.

Funding and support from the U.S. National Science Foundation, California Air Resources Board, National Park Service, U.S. Environmental Protection Agency, U.S. Forest Service, U.S. Geological Survey, University of California, and National Aeronautics and Space Administration are gratefully acknowledged.

ONE

Introduction to the Sierra Nevada

Abstract. Physical, geographic, and ecological features and cultural aspects of the Sierra Nevada are described as an introduction to the Sierra Nevada, John Muir's "range of light." The range extends about 700 km north to south and reaches an elevation of 4,421 m. Pleistocene glaciations, last retreating between 10,000 and 25,000 years ago, sculpted valleys and high- elevation lake basins. Native Americans lived on the flanks of the Sierra and utilized its rich ecological diversity. Beginning with the influx of Europeans sparked by the discovery of gold, the natural resources in the high-elevation Sierra have been exploited for mining, grazing, water supply, and hydroelectric power. The Sierra Club, founded in 1892, has played a key role in fostering recreational uses and conservation of the region. Subsequent legislative actions have led to the establishment and expansion of national parks and national forests throughout the Sierra.

Key Words. physical features, geographic features, ecological features, natural resources, conservation, Native Americans

THE SIERRA NEVADA OF CALIFORNIA extends about 700 km (40°15′N to 35°N), with a width of 100 km to 130 km, and reaches an elevation along its crest at Mount Whitney of 4,421 m (figures 1 and 2). To one group of Native Americans it was known as "rock placed on rock." A Franciscan missionary called it "una gran sierra nevada"—a great snow-covered range. To John Muir, the Sierra Nevada was the "range of light." Scenes of the region are widely known from the iconic photographs by Ansel Adams (e.g., *Moon and Half Dome*, 1960; *Clearing Winter Storm*, 1940; *Winter Sunrise*, 1944), his collected works (e.g., Adams 1979), and the Sierra Club photo-essay,

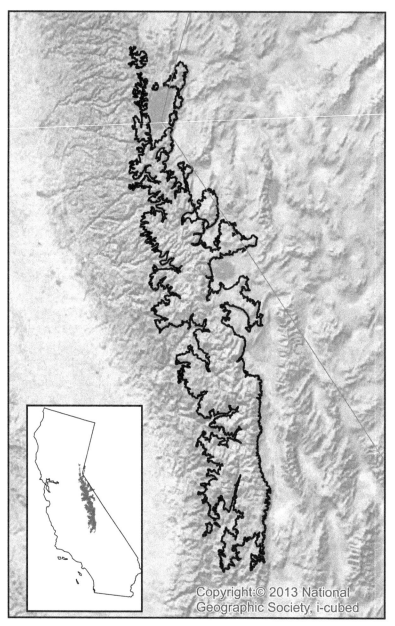

FIGURE 1. Sierra Nevada location and topography. Environmental Systems Research Institute (ESRI) base map is shaded relief using National Geographic Style (www.esri .com/en-us/arcgis/products/data-location-services/data/basemaps-imagery#image7). Solid black line is 2,300 m contour.

FIGURE 2. High-elevation Sierra landscapes and lakes. Photograph by S. Sadro.

Gentle Wilderness (Kaufman 1964). Indeed, the Sierra Nevada is all of these and more, as its cultural history and scientific discoveries have revealed.

This book provides a synthesis of our current understanding of the hydrology, watershed biogeochemistry, and limnology of high-elevation lakes and their watersheds and the impacts of environmental changes on these aspects. For our analysis, we define "high elevation" as lakes above about 2,300 m. We combine three decades of measurements from specific watersheds with regional surveys, experimental results, and paleoecological data. Acidification, eutrophication, invasive species, and climate change are environmental issues addressed in multiple facets. As an introduction we begin with a description of the physical, geographic, and ecological features and cultural aspects of the Sierra Nevada.

PHYSICAL FEATURES

The Sierra Nevada has evolved over the past 500 million years (Hill 1975) and is composed largely of igneous and metamorphic rocks of diverse composition and age. By about 70 million years ago (Mya), the granitic rocks forming the Sierra batholith were in place. From 80 Mya to 40 Mya much of the metamorphosed rock deposited in the Paleozoic and Mesozoic eras was removed by erosion. During the mid- to late Tertiary period basic volcanic material buried much of the northern Sierra, and volcanism has continued.

The traditional view suggests that increased tectonic activity in the past 3 million to 5 million years is responsible for the current elevation. However, some recent studies provide evidence that the Sierra Nevada was close to its current elevation 40 Mya to 50 Mya and may have been the edge of a pre-Eocene continental plateau (Mulch, Graham, and Chamberlain 2006; Mulch et al. 2008; Crowley, Koch, and Davis 2008). Pleistocene glaciations occurred several times over the past 3 million years, last retreating between 10,000 and 25,000 years ago and leaving sculpted valleys and high-elevation lake basins.

The sloping western side of the Sierra contrasts with the steep escarpment on the eastern flank and reflects the broad genesis of the range as a tilted faulted block that continues to uplift. The elevation of the Sierra crest and the height of peaks tend to increase from Desolation Wilderness in the north to the region around Mount Whitney in the south, where thirteen peaks reach above 4,200 m (Storer, Usinger, and Lukas 2004). The high-elevation Great Western Divide in Sequoia National Park separates the Kern River from the Kings and Kaweah watersheds. Steep-sided and deep valleys dissect the eastern and western sides of the range, with Yosemite Valley and the Kings and Kern Canyons being the most spectacular.

Sierra soils derived primarily from granites tend to be slightly acidic (pH 5–6.5), with limited capacity to retain moisture compared to soils formed from volcanic and metamorphic rocks (Melack and Stoddard 1991). Soils derived from volcanic and metavolcanic rocks are finer textured and have more iron- and magnesium-rich minerals and calcic feldspars than soils derived from granites. High-elevation soils (most commonly inceptisols) experience cold temperature regimes, with mean annual temperatures ranging from 0°C to 8°C.

Warm, dry summers and cool, wet winters are characteristic of the Sierra Nevada. Typically, most precipitation occurs in winter as snow, with only occasional summer rain. Statewide, California has high year-to-year variability in precipitation (Dettinger, Redmond, and Cayan 2004). Prevailing westerly winds entering California are strongly influenced by the position of the North Pacific high-pressure cell (Iacobellis et al. 2016). During summer, the high-pressure system is largest and strongest, and its descending air suppresses rainfall. As the high pressure weakens and shifts to the southeast in the winter, precipitation-bearing storms enter California from the Pacific. Features called atmospheric rivers, bands of air with large amounts of moisture from the tropical or subtropical Pacific Ocean, are responsible for the majority of the winter rain and snow falling in the Sierra Nevada (Dettinger et al. 2011). During summer, infrequent convective storms, sometimes associated with the North American monsoon, can occur.

Climatic variation associated with the El Niño/Southern Oscillation (ENSO) is strong in the Sierra Nevada. During El Niño events heavy snowfall often occurs, but about one-third of El Niño winters are dry. In contrast, during the La Niña phase, precipitation is usually below average (Redmond and Koch 1991). Large differences in orographically influenced precipitation occur between the western and eastern sides of the Sierra Nevada (Pandey, Cayan, and Georgakakos 1999). Analyses of the regional and temporal differences in precipitation and their consequences are considered in subsequent chapters. Two of the best records of meteorological measurements for the high-elevation Sierra Nevada are from our studies in the Tokopah basin, in Sequoia National Park, and at Mammoth Mountain. These data are utilized in our examination of hydrological and limnological aspects of Sierra lakes and watersheds in later chapters.

ECOLOGICAL ASPECTS

The natural history of the flora and fauna of the Sierra is described by Whitney (1979), Storer, Usinger, and Lukas (2004), Smith (1976), and the High Sierra hiking guides (e.g., Felzer 1981; Selters 1980).

Vegetation in the Sierra Nevada is controlled by a combination of temperature, precipitation, and soils, and there are three broad vegetation zones (Rundel, Parsons, and Gordon 1977; Majors and Taylor 1977). The elevational ranges of these zones rise from north to south and are higher east compared to west of the crest. All three zones are characterized by deep winter snow and infrequent summer rain. The upper montane zone occurs at elevations of approximately 2,100 to 2,700 m and is often characterized by nearly pure stands of red fir and lodgepole pine. Jeffrey pine and western juniper are also found in this zone, particularly on drier slopes. The growing season is 12 to 16 weeks, with 40 to 70 frost-free days annually. The subalpine zone extends from approximately 2,700 to 3,100 m and is relatively sparsely forested. Dominant tree species include whitebark pine, foxtail pine (southern Sierra), mountain hemlock (central and northern Sierra), lodgepole pine, and western white pine. The growing season is 6 to 9 weeks, and frosts can occur during all summer months. The alpine zone occurs at the highest elevations (> 3,100 m), and harsh climatic conditions prevent tree establishment. Vegetation is sparse and is composed of short-statured shrubs that grow in the limited areas containing soil. The growing season is short, and temperatures are cool even during summer. The tree line occurs at between 3,200 and 3,400 m, and above about 3,200 m vegetation is largely only alpine meadows, often growing near lakes.

A variety of fungi, lichens, ferns, and mosses are scattered through the high-elevation Sierra, including the giant lenticus (*Nelentinus ponderosus*) living on decaying conifers; the brown tile-lichen (*Lecidea atrobrunnea*), which imparts a gray cast to the granite; and fragile fern (*Cystopteris fragilis*) and star moss (*Syntrichia ruralis*). Plants such as sky pilot (*Polemonium eximium*) occur above 3,500 m. Other plants living at high elevations include western wallflower (*Erysimum capitatum*), alpine gold (*Hulsea algida*), lupine (*Lupinus* spp.), asters (*Aster* spp.), daisies (*Erigeron* spp.), mountain sorrel (*Oxyria digyna*), alpine columbine (*Aquilegia pubescens*), Sierra shooting star (*Dodecatheon jeffreyi*), and alpine phlox (*Phlox* spp.). Common shrubs include white heather (*Cassiope mertensiana*), red mountain heather (*Phyllodoce breweri*), alpine laurel (*Kalmia polifolia*), wax currant (*Ribes cereum*), alpine gooseberry (*Ribes montigenum*), and mountain ash (*Sorbus californica*).

At upper elevations common mammals include pika (*Ochotona princeps*), yellow-bellied marmot (*Marmota flaviventris*), Belding ground squirrel (*Spermophilus beldingi*), heather vole (*Phenacomys intermedius*), alpine chipmunk (*Tamias alpinus*), and occasional black bear (*Ursus americanus*) and coyote (*Canis latrans*). Predators include long-tailed and short-tailed weasels (*Mustela frenata* and *M. erminea*), marten and fisher (*Martes americana* and *Martes pennanti*), gray fox (*Urocyon cinereoargenteus*), and the threatened Sierra Nevada red fox (*Vulpes vulpes necator*). Grizzly bears (*Ursus arctos*) have been extirpated from the region. California bighorn sheep (*Ovis canadensis*) are now endangered. Reptiles are limited to the western terrestrial garter snake and the Sierra garter snake (*Thamnophis elegans* and *T. couchii*). Rattlesnakes (usually *Crotalus oreganus*) seldom venture above 2,300 m, but rare sightings have been reported on south-facing slopes at 3,000 m. Amphibians include the Mount Lyell salamander (*Hydromantes platycephalus*), the endangered Yosemite toad (*Bufo canorus*), Pacific treefrog (*Hyla regilla*), and the endangered mountain yellow-legged frog (*Rana muscosa*, *Rana sierrae*). Examples of birds commonly observed in the High Sierra include the mountain chickadee (*Poecile gambeli*), red-tailed hawk (*Buteo jamaicensis*), yellow-rumped warbler (*Dendroica coronate*), rock wren (*Salpinctes obsoletus*), white-throated swift (*Aeronautes saxatalis*), common raven (*Corvus corax*), water ouzel (*Cinclus mexicanus*), gray-crowned rosy finch (*Leucosticte tephrocotis*), Clark's nutcracker (*Nucifraga columbiana*), American robin (*Turdus migratorius*), mountain blue bird (*Sialia currucoides*), hermit thrush (*Catharus guttatus*), and dark-eyed junco (*Junco hyemalis*).

CULTURAL HISTORY

Not much is known about Native Americans who lived in the Sierra Nevada before the arrival of Europeans. The northern, central, and southern Sierra Miwoks occupied the foothills and mountains along the central western Sierra, with populations of each estimated to be from approximately 2,000 to 2,700 (Levy 1978). As was typical of other Native Americans living in the Sierra, the Miwok hunted black and grizzly bear, deer, smaller mammals, and game birds and fished and harvested a variety of plant products. They traded for salt and obsidian with the Washoe and Paiute living in the Great Basin. The Monache (also called the western Mono) were composed of six tribal groups along the western Sierra, and they also crossed the Sierra for trading (Spier 1978). In the southern Sierra along the Kern River, the Tubatulabal lived much as their neighbors to the north (Smith 1978). On the eastern flanks of the Sierra, the Owens Valley and northern Paiute and the Washoe made a living (Liljeblad and Fowler 1986; Fowler and Liljeblad 1986; D'Azevedo 1986). The Owens Valley Paiute occupied semipermanent camps with artificial irrigation of native plants and numbered between 1,000 and 2,000 before white settlement. Obsidian fragments are found throughout the lower elevations of the east side of the Sierra Nevada, even though obsidian outcrops in only a few locations. The northern Paiute ranged from Mono Lake into southern Oregon, and the Washoe lived around Lake Tahoe and numbered about 1,500. Trading and some intermarriage appear to have occurred along and across the Sierra among at least some of these groups. Smallpox, scarlet fever, measles, and violence caused by European settlers led to large population declines and the cultural demise of the Native Americans by the mid- to late 1800s in much of the Sierra Nevada.

The first Europeans to traverse the Sierra included the trapper Jedediah Smith in 1827, Joseph Walker in 1833, and John C. Frémont in the mid-1840s (Farquahar 1965). The discovery of gold, the so-called Mother Lode, on the Sierra's western slopes in 1848 precipitated the first mass migration to California, and by the mid-1860s the nonindigenous population increased sharply to about 223,000. With support from the California State Legislature, the so-called Whitney Survey began in 1860 and reached into remote recesses of the Sierra. Over subsequent decades the population of California continued to increase and by the 2010 census had reached slightly more than 38 million, with about 70% living in coastal counties and over 90% in urban areas (Alagona et al. 2016).

HISTORY OF NATURAL RESOURCE USE AND CONSERVATION

Beginning with the influx of Europeans sparked by the discovery of gold in California, the natural resources in the high-elevation Sierra Nevada have been exploited for mining, grazing, water supply, and hydroelectric power. Fish were introduced to provide food for miners and more recently to support sportfishing (Christenson 1977; Knapp 1996). Starting during the Civil War to supply wool, sheepherding in Sierra meadows increased through the late nineteenth century. Grazing by sheep, called "hoofed locusts" by John Muir, and the fires set by the ranchers caused substantial damage to the vegetation and soil. Human alteration had occurred earlier with burning by Native American peoples to maintain meadows and remove understory growth to enhance hunting and travel (Gruell 2001). A well-referenced historical account of the European exploration and utilization of the Sierra Nevada is provided by Farquhar (1965). Natural resource values and management issues in the Sierra Nevada forests and surroundings are treated extensively by the Sierra Nevada Ecosystems Project (1996) report.

Placer gravels on the western slope contain gold, and the hydraulic mining that was used to expose these gravels deformed the stream channels and led to downstream movement of vast amounts of debris. Tons of mercury used during the gold rush to amalgamate gold have created a legacy of aquatic food web contamination in the Sacramento–San Joaquin Bay-Delta (Singer et al. 2013). Quartz veins in metamorphic bands also contained gold, and these deposits led to further mining activities, as did discovery of deposits of silver, especially the Comstock Lode near Lake Tahoe, and of copper, lead, zinc, chromium, and tungsten (Sharp 1976). For example, the Copper Mountain Mining Company was active beginning in the 1890s in the remote Deadman and Cloud Canyons in the Triple Divide Area that became part of Sequoia National Park in 1926 (Selters 1980). Though mining in the Mineral King area of the southern Sierra started in the 1870s, it was not sufficiently productive and largely ended by the 1880s (Felzer 1981).

Starting in 1868, John Muir, a Scottish sheepherder, began his wandering and writings about the Sierra. The Sierra Club was founded by John Muir in 1892 and has played a key role in fostering recreational uses and conservation of the region. In 1864 the U.S. Congress ceded Yosemite Valley to the State of California as a park; striking photographs by Carleton Watkins taken in the 1800s played a key role in the establishment of the park (Green 2018). Subsequent legislative actions have led to the establishment and expansion of national parks and national forests and, more recently, wilderness designations within them. Currently, about 1.5 million hectares of the Sierra lie within national parks and national forests. Hence most human activities are now recreational hiking, camping, and fishing, and lakes and streams are an important focus of these activities. Between 4 million and 5 million people visit the area each year.

In popular culture, the Sierra Nevada has been used as the backdrop for many Hollywood films beginning with the silent era and continuing to the present day. Prominent locations include the Alabama Hills and Whitney Portal near Lone Pine and the Lake Tahoe and Truckee area (Lone Pine Film History Museum, Truckee-Donner Historical Society). Interestingly, one of our main study sites, Pear Lake, was prominently featured in the 2003 blockbuster film *Hulk*, directed by Ang Lee.

While several dams have created or modified lakes at high elevations (e.g., Sabrina, South, Edison), most large and moderate-sized reservoirs are located at lower elevations on the western slopes of the Sierra Nevada. These reservoirs are operated for water supply, irrigation, and consumption, as well as for flood control, recreation, and generation of electricity. The California Water Plan (Department of Water Resources 2009) provides an overview of the network of major reservoirs and water conveyance systems throughout California. For example, planning to import water from the Tuolumne River followed the 1906 earthquake in San Francisco, and Hetch Hetchy Valley, located within Yosemite National Park, was identified as a potential dam site. Despite opposition from John Muir and the Sierra Club, O'Shaughnessy Dam, creating Hetch Hetchy Reservoir, was completed in 1923, after Congress passed the Raker Act in 1913 (Hundley 2001).

Though remote from most direct damage, the watersheds and lakes of the high-elevation Sierra are susceptible to purposeful or accidental introductions of alien species, atmospheric transport of contaminants, and regional climate changes, as explored in subsequent chapters.

TWO

Introduction to High-Elevation Lakes and Watersheds of the Sierra Nevada

Abstract. The Sierra Nevada is a region with thousands of high-elevation lakes and ponds. The abundance of Sierra lakes is highest in the central part of the range and tapers at both the northern and southern ends in conjunction with changes in elevation, precipitation, and underlying geology. Although lakes distributed throughout the Sierra have been sampled to different degrees, the most extensive, intensive, and longest record comes from Emerald Lake and its watershed in Sequoia National Park. Complementary measurements are available for three watersheds located on the eastern side (Ruby, Crystal, Spuller), one at the northern end (Lost), and three on the southwestern side (Pear, Topaz, and the upper Marble Fork of the Kaweah River). Each of these lakes and watersheds is briefly described.

Key Words. Emerald Lake, Ruby Lake, Crystal Lake, Spuller Lake, Pear Lake, Topaz Lake, Lost Lake

LAKES OF THE SIERRA NEVADA

The Sierra Nevada is a lake-rich region, with iconic Lake Tahoe and scenic and saline Mono Lake (reviewed in Melack and Schladow 2016; Patten et al. 1987; Melack et al. 2017), in addition to thousands of other high-elevation lakes as well as ponds. The number of lakes and ponds within the glacier-carved high-elevation basins of the Sierra Nevada is remarkable; there are at least 12,000 water bodies located above an elevation of 2,280 m along the length of the range. To determine the abundance and distribution of lakes throughout the Sierra Nevada, we created a comprehensive data set by combining information from multiple data sources. California Department of Fish and Wildlife (CDFW) data were augmented with data collected from Yosemite National Park, Sequoia and Kings Canyon National Parks, and surveys by Roland Knapp (Sierra Nevada Aquatic Research Laboratory, University of California, Santa Barbara). We used digital elevation models (1 arc-second, ~30 m) from the National Elevation Database (www.usgs.gov/core-science-systems/national-geospatial-program/national-map) to determine the elevation of each lake and the average slope for a 250 m band around the lakes. Land cover within the 250 m band was determined using the landscape classifications in the National Land Cover Database (2001, www.mrlc.gov; 1 arc-second resolution).

The distribution of lakes in the Sierra Nevada is largely a result of glacial activity. Sierra lakes are most abundant in the central part of the range and taper in number at both the northern and southern ends in conjunction with changes in elevation, precipitation, and underlying geology (figures 3 and 4). The longitudinal distribution of lakes is skewed toward the east, with abundances peaking just west of the Sierra crest. This east-west distribution is related to elevation, which is typically higher near the eastern crest, where glacial scouring and deposition occurred more frequently at higher elevations, causing an increase in lake abundance at around 3,300 m. Glaciation did not extend to the uppermost elevations (Moore and Mack 2008; Moore and Moring 2013).

The majority of high-elevation water bodies are small; nearly three-fourths of them might be considered ponds based on their surface area (< 0.5 ha) (figure 5). At least a third of these ponds are classified as ephemeral in the CDFW database, holding water only through spring snowmelt and the first part of summer in most years. The 3,151 water bodies with areas > 0.5 ha might be considered lakes, and all are perennial (figure 6). Of these, nineteen are reservoirs, which comprise seven of the ten largest water bodies. Unlike the large drowned river valley reservoirs found within the western foothills of the Sierra, these high-elevation reservoirs are more often natural lakes whose outlets have been dammed to increase storage capacity.

While ponds dominate the Sierra in terms of abundance, they collectively account for less than 7% of the cumulative surface area. Ponds are surprisingly regular in size; median surface area and interquartile range, which characterizes dispersion between the 25th and 75th quartiles, are both 0.1 ha. In contrast, median lake area, excluding reservoirs, is in the 1.8 ha ± 3.3 ha interquartile range. The higher dispersion is due to the abundance of relatively small lakes. The reservoirs in our study were similarly skewed; they had

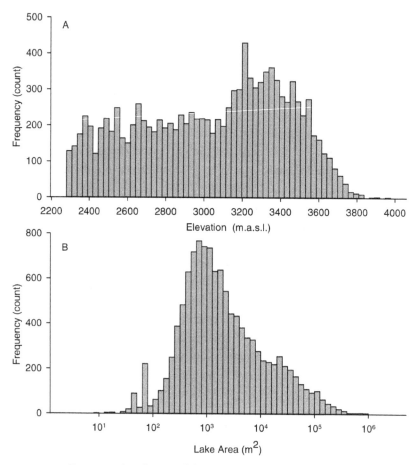

FIGURE 3. Frequency distribution of the ~12,000 lakes and ponds (smallest ~10 m²) found above an elevation of 2,300 m sorted by altitude (A) and size (B).

an average area of 141 ha, which is about twice the median area of 71 ha ± 100.4 ha interquartile range. All lakes, including reservoirs, collectively account for 93% of the surface area. Reservoirs alone constitute 16% of the total surface area of all lakes and ponds.

Although we have data on maximum depth for only a subset of lakes and ponds (n = 8712), their area distribution matches that of the entire lake distribution, suggesting that values are broadly representative for the Sierra Nevada. Because of their small and consistent size, ponds have the shallowest and most uniform depths. Median pond depth was 1.0 m ± 1.2 m interquartile range, which was about the same as their 1.2 m mean depth. In contrast,

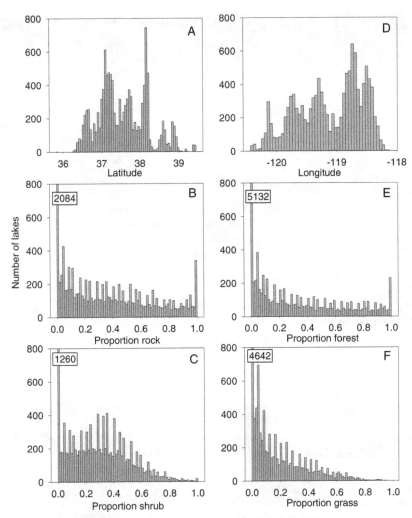

FIGURE 4. Frequency distributions of the ~12,000 lakes and ponds (smallest ~10 m²) found above an elevation of 2,300 m sorted by latitude (A) and longitude (D) across the Sierra. Frequency distributions for the land cover attributes in the 250 m radius surrounding lakes as proportions: rock (B), shrubs (C), forest (E), and grass (F).

median depth of lakes was 5.7 m ± 7.3 m interquartile range. However, this is likely to be a slight underestimate due to constraints on the type of gear used to measure maximum depths in some lakes. The median depth and interquartile range for reservoirs is 11.1 m ± 15.6 m, indicating that high-elevation reservoirs are largely natural lakes whose levels have been artificially raised by small dams. The deepest lakes in our data set are 80 to 85 m.

FIGURE 5. Tarn above Gem Lake, Rock Creek basin, eastern Sierra. Photograph by S. Sadro.

FIGURE 6. Big Bird Lake, western Sierra. Photograph by S. Sadro.

In addition to lake area and depth, lake shape and total volume have important implications for sedimentation dynamics, thermal stratification, and physical processes such as mixing. Lake morphology in the Sierra is fairly consistent. This likely reflects the similar geologic processes that carved the lakes during the last glacial period and similar weathering processes. Bathymetries of lakes, and the depth-volume and depth-area hypsographic curves, which represent graphically the relationships that structure the shape of the basin, reflect the bowl-shaped basins of most lakes. Even relatively oblong lakes tend to have simple basin geometries wherein the deepest part of the lake lies along the longitudinal axis. Lake bottoms typically descend rather steeply from the shore, limiting the extent of shallow littoral areas. When present, littoral zones are often found near inlets because of deposition of sediments. A small number of lakes do have complex morphometries, including islands and multiple basins.

Our estimates of lake volume were made using a regression model on the basis of bathymetries measured for seventeen lakes in conjunction with maximum depth and lake area and perimeter. We approximated the volume of the lake as half of an ellipsoid, where each lake surface was represented as an ellipse, with its semi-major and semi-minor axes determined from the area and perimeter, and the maximum depth as the third axis of the ellipsoid. The volumes of the half-ellipsoid were about half the measured volumes, but the statistical relationship is robust ($R^2 = 0.99$, $F_{15,2} = 449$, $p < 0.0001$, Volume = $1.921 \times EllipsoidVolume^{0.996}$). This model is not biased, with the residuals (as fractions of lake size) independent of the size of the lakes. Although we use this model to estimate volumes for ponds as well, the accuracy of the estimates has not been tested with measured volumes based on a suite of bathymetric measurements over a larger set of lakes.

A defining feature of the Sierra Nevada is the presence of numerous gradients associated with elevation. Soil development, the extent and type of vegetation present, and temperature are all correlated with elevation. The distribution of lakes along these gradients has important implications for the physical and biogeochemical processes that take place within them (Sadro et al. 2012), which in turn establish the context for important ecosystem functions. Because the area immediately surrounding a lake can have as strong an influence on biogeochemistry as cumulative catchment influences, we examined the landscape characteristics in a 250 m buffer surrounding each lake and pond in our database. Median rock cover surrounding lakes was 27% ± 52% interquartile range, median forest cover was 5% ± 37% interquartile range, and median shrub cover was 28% ± 33% interquartile range. The average slope surrounding lakes was up to 43 degrees and had a median value of 12.5 degrees ± 9.2 degrees interquartile range.

Water bodies in the Sierra can be classified into groups based on lake and watershed characteristics. A hierarchical cluster analysis identified four primary groups. Water bodies were primarily characterized by elevation-related changes in landscape attributes, to a lesser extent by size. The four broad classifications are (1) lower-elevation forested ponds; (2) medium- to higher-elevation lakes and ponds with variable land cover; (3) higher-elevation lakes and ponds mostly surrounded by rock; and (4) higher-elevation lakes and ponds mostly surrounded by meadow and shrubs. We report median and interquartile range values.

The first group, with 27% of all water bodies, consists mostly of lower-elevation (2,698 ± 352 m.a.s.l.) ponds with small surface areas (0.1 ha ± 0.4 ha) that tended to be surrounded primarily by forest (64% ± 41%) and shrub (22% ± 25%), with some adjacent rock (4% ± 15%) but negligible grass cover. The second group, with 30% of all water bodies, spans a wide range in elevation (2,914 ± 622 m.a.s.l.) and consists of lakes and ponds that tend to be slightly larger in size (0.2 ± 1.0 ha) than the first group. Water bodies in the second group were surrounded primarily by rock (26% ± 37%) and shrub (48% ± 22%), with thin forest cover (9% ± 21%) and some grassy meadow areas (3% ± 12%). The third group, consisting of 20% of all water bodies, was made up mostly of high-elevation lakes and ponds (3,437 ± 270 m.a.s.l.) that are comparatively small (0.1 ha ± 0.4 ha). Located primarily above the tree line, this group's dominant land cover was rock (83% ± 25%), with thin shrub cover (6% ± 19%) and some grassy areas (4% ± 11%). The fourth group comprises 22% of all lakes. It consists mostly of high-elevation lakes and ponds (3,314 ± 256 m.a.s.l.) with a median and interquartile size range of 0.1 ha ± 0.5 ha. These water bodies were located above the tree line in rocky areas (3% ± 37%), but unlike the first group, they tended to be more frequently surrounded by grassy meadows (39% ± 22%) and shrubs (25% ± 24%). Two larger reservoirs at the lower limit of the elevation range clustered together based on their substantially larger size.

DESCRIPTIONS OF SPECIFIC LAKES AND WATERSHEDS

Since the early 1980s, high-elevation lakes and watersheds in the Sierra Nevada have received increased attention, motivated in part by concern about acidic atmospheric deposition. Initially, surveys of lake chemistry were

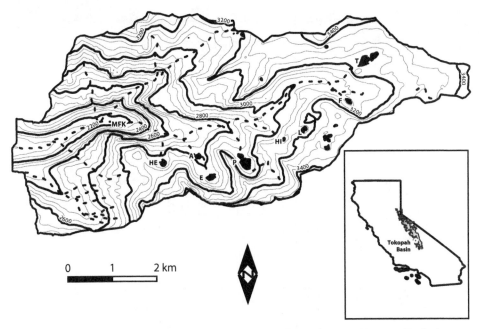

FIGURE 7. Tokopah basin (upper Marble Fork of the Kaweah River), Sequoia National Park, showing topography, lakes, and streams. Elevation contours are in meters. Site names are abbreviated as follows: HE - Heather Lake, A - Aster Lake, E - Emerald Lake, P - Pear Lake, HI - Hidden Lake, L - Lyness Lake, F - Frog Lake, T - Topaz Lake. Derived from U.S. Geological Survey, 2017, 1/3rd arc-second Digital Elevation Models (DEMs), USGS National Map 3DEP downloadable data collection, www.usgs.gov /core-science-systems/ngp/3dep.

conducted to determine the geographic extent of sensitive lakes (Melack, Stoddart, and Ochs 1985; Landers et al. 1987; Melack and Stoddard 1991). Concurrently, the California Air Resources Board initiated a study of the Emerald Lake watershed that, in addition to investigating atmospheric deposition, included research on the terrestrial watershed and aquatic ecosystems and added a larger set of Sierra lakes and watersheds. Here we characterize several catchments that will be especially relevant in subsequent chapters. The eight sites span the Sierra Nevada, with three watersheds located on the eastern side (Ruby, Crystal, Spuller), one located at the northern end (Lost), and the remainder located at the southwestern side (Emerald, Pear, Topaz, and the upper Marble Fork of the Kaweah River, also called the Tokopah basin) (table 1). Seven of the catchments are glacial cirques with lakes. The upper Marble Fork drains the Tokopah basin of Sequoia National Park, a glacially carved basin of about 2,000 ha that includes within its boundaries

TABLE 1 Summary of specific lake and watershed characteristics

Lake/Basin	Basin Area (ha)	Basin Relief (m)	Lake Area (ha)	Outlet Elev. (m)	Lake Max. Depth (m)	Lake Mean Depth (m)	Lake Volume (m³)	Lake Volume to Watershed Area Index (m³/m²)
Crystal	135	293	5.0	2,951	14	6.5	324,000	0.24
Emerald	120	616	2.7	2,800	10	6.0	162,000	0.14
Lost	25	160	0.7	2,475	5.5	1.9	12,500	0.05
Pear	136	471	8.0	2,904	27	7.4	591,000	0.43
Ruby	441	812	12.6	3,390	35	16.4	2,080,000	0.47
Spuller	97	537	2.2	3,131	5.5	1.6	34,700	0.04
Topaz	178	275	5.2	3,218	5.0	1.5	76,900	0.04
Tokopah basin	1,908	872	30	2,621	NA	NA	1,059,000	0.06

NOTE: Basin area is the amount of drainage area above the outflow gauging station. The lake volume to watershed index is the volume of a lake (m³) divided by the area of the lake's drainage basin (m²). For the Tokopah basin, lake volume is the sum of the volumes of Emerald, Pear, and Topaz Lakes plus an additional 230,000 m³ estimated lake volume for Aster Lake and other small ponds in the catchment.

the watersheds for Emerald, Pear, and Topaz Lakes, along with several other small lakes and ponds (figure 7). A wide suite of environmental and limnological variables have been measured in these lakes, with the most extensive, intensive, and longest in Emerald Lake and its watershed.

Lost Lake

Lost Lake (38°51′37″N, 120°5′48″W) is located near Lake Tahoe in the Desolation Wilderness of Eldorado National Forest. Thermal stratification occurs during the summer but is weaker than winter stratification when near anoxia (dissolved oxygen ~ 0 mg L^{-1}) develops in the waters below thermocline. Surface waters had dissolved oxygen levels above 5 mg L^{-1} during other periods. Maximum lake temperature ranged from 13°C (1993) to 21°C (1991). A population of non-native brook trout is present. The Lost Lake watershed has a north-facing aspect. Hemlock, lodgepole pine, and western white pine line the shore of the lake and are scattered, along with shrubs, throughout the watershed; wet meadows are found around the lake. Most of the catchment is composed of bedrock or bedrock with patches of grass-covered, alpine brown soils, based on a scheme described by Melack et al. (1998).

Spuller Lake

Spuller Lake (37°56′55″N, 119°17′2″W) is located in the Hall Research Natural Area immediately east of Yosemite National Park. It does not develop persistent seasonal stratification because its shallow depth allows frequent mixing. In the winter the lake was stratified, and dissolved oxygen depletion occurred near the bottom. At other times of the year, lake waters were well oxygenated. The watershed has a northeast aspect and is nearly devoid of trees. Vegetation is confined to areas near the lake as meadows of grass and sedges and small stands of dwarfed and stunted whitebark pine. A small population of reproducing non-native brook trout was present at least through the mid-1990s. Most of the watershed is talus and bedrock outcrops composed of medium-grained hornblende-biotite granodiorite with associated alpine brown soils (Bateman et al. 1988).

Crystal Lake

Crystal Lake (37°35′36″N, 119°01′05″W) is located about 10 km southwest of the town of Mammoth Lakes. Maximum summertime surface temperatures in the lake ranged from 12°C to 15°C. Surface waters of the lake were well oxygenated (dissolved oxygen > 7 mg L^{-1}); however, dissolved oxygen concentrations of less than 5 mg L^{-1} at depth were common during the winter months. Crystal Lake contains non-native brook and rainbow trout. The watershed has a north-facing aspect and is sparsely forested for about half its area with a mixture of whitebark and lodgepole pine. Much of the runoff from the catchment flows through a meadow of heather, grasses, and sedges and other shrubs. The eastern and southern portions of the basin are dominated by a granitic dome and extensive talus; the remaining basin is dominated by bedrock and soils of volcanic origin (Huber and Rinehart 1965). Most soils are volcanic brown soils, occurring in various assemblages of rock, scree, and talus; in the inlet meadow volcanic wet meadow soils occur.

Ruby Lake

Ruby Lake (37°24'50"N, 118°46'15"W) is situated near the eastern edge of the Sierra Nevada crest in the John Muir Wilderness of Inyo National Forest. Strong to moderate thermal stratification occurred during both winter and summer, and intervals of low deep-water oxygen were observed. Maximum summer lake temperature ranged from 11°C to 16°C. Sparse stands of whitebark and lodgepole pine are confined to areas along the north and northeast edge of the lake, and most vegetation is grasses, sedges, and low shrubs. The basin has a northwestern exposure, and the bedrock is predominantly quartz monzonite (Lockwood and Lydon 1975). Bedrock outcrops, talus, and boulders cover most of the catchment, and soils are sparse. Rock glaciers in the watershed result in glacial flour in the major inflow to the lake.

Emerald Lake

Emerald Lake (36°35'49"N, 118°40'29"W) is located in Sequoia National Park in the Triple Divide Area of the Sierra Nevada. Thermal stratification is weak during the summer; inverse stratification occurs during winter and spring. Peak summer temperatures in the lake ranged from 11°C to 20°C. Surface waters in Emerald Lake are well oxygenated year-round; periods of low oxygen at depth have been observed during both winter and summer stratification. A reproducing population of non-native brook trout is found in the lake. The basin has a northern exposure. Vegetation in the Emerald Lake basin is sparse (Rundel, St. John, and Berry 1988), consisting of scattered conifers (lodgepole and western white pine), low woody shrubs, grasses, and sedges. Bedrock is mainly granodiorite with mafic inclusions, aplite dikes, and pegamite veins (Clow 1987; Sisson and Moore 1984). Poorly developed soils cover about 20% of the watershed, and these are slightly acidic and weakly buffered; the primary clay minerals are vermiculite, kaolinite, and gibbsite (Huntington and Akeson 1987; Lund and Brown 1989).

Pear Lake

Pear Lake (36°36′02″N, 118°40′00″W) is located in Sequoia National Park about 1 km northeast of Emerald Lake. Thermal stratification is strong during much of the year, and low dissolved oxygen concentrations occur at depth. Mixing of the lake occurs during spring and autumn, but spring mixing can be incomplete in some years; water temperatures at a depth of 25 m rarely exceeded 5°C. The persistence of deep water summer anoxia observed in Pear Lake is unusual among the other lakes we sampled and likely influenced by its bathymetry, which is characterized by shallow to moderate depths around the lake perimeter and a small deep depression (25–27 m) along the southern lakeshore. What little vegetation is found in the basin consists of a few stands of coniferous trees (lodgepole pine, western white pine, red fir), shrubs, grasses, and sedges. Soils, where present, are classified as alpine brown. Most of the Pear Lake watershed is composed of coarse-grained granites containing sparse mafic inclusions of widely variable size and texture. The remainder of the basin is underlain by medium-grained, porphyritic granodiorite (Sisson and Moore 1984). Greater than 90% of the catchment is composed of bedrock, talus, and boulders.

Topaz Lake

Topaz Lake (36°37′30″N, 118°38′11″W) is located at the head of the Tokopah basin in Sequoia National Park, about 6 km north-northwest of Emerald Lake. The lake is connected by a network of narrow channels to a small, shallow pond during high-water periods. In winters with especially large snowfall, such as 1983, it is likely that much of the lake is composed of ice and slush. Seasonal thermal stratification of the lake is confined to the winter months. Low dissolved oxygen concentrations were measured during most winters; surface waters were well oxygenated. Maximum summertime lake temperatures ranged from 14°C to 19°C. No fish were observed in Topaz Lake, despite the fact that it was stocked with trout several times in this century. Trout exert substantial influence on the zooplankton and zoobenthos of Sierra lakes (Stoddard 1987; Melack et al. 1989). In contrast to lakes with fish, large zooplankters (*Hesperodiaptomus eiseni, Daphnia melanica*) are plentiful in Topaz Lake (Melack et al. 1993). The watershed has a southern exposure, and parts of the upper basin have extensive meadows (grasses and sedges), and a small stand of foxtail pines occurs in the upper eastern portion. Extensive wet meadows are found along the north shore of the lake and around the ponds in the basin. Alpine brown soils are found throughout the watershed, often forming complexes with rocks and bedrock outcrops. The geology of the basin is dominated by fine-grained, porphyritic granodiorite containing abundant mafic inclusions (Sisson and Moore 1987).

Upper Marble Fork of the Kaweah River (Tokopah Basin)

The upper Marble Fork of the Kaweah River, or Tokopah basin (36°36'22"N, 118°40'59"W), is located in Sequoia National Park. The Tokopah basin includes Emerald, Pear, and Topaz Lakes, which, along with several other small ponds and lakes, make up 30 ha of the basin's 1,908 ha drainage area. Non-native brook trout have been observed but appear to be restricted to reaches below the confluence of the Pear Lake outlet. Trees, mainly lodgepole pine and western white pine and willows, are found along the river as it meanders through the basin. A grove of Sierra junipers (*Juniperus occidentalis australis*) grows near the site of our Marble Fork gauging station. The higher elevations have sparse vegetation composed primarily of sedges and grasses. The geology of the Tokopah basin is dominated by fine- and medium-grained porphyritic granodiorite and coarse-grained granite. Most of the basin is composed of bedrock and talus, but there are significant areas of wet meadow soils in upper portions of the basin and along the margins of the river.

THREE

Snow Hydrology

Abstract. In the high-elevation Sierra Nevada, most of the precipitation falls as snow from November to April, and the snow melts from April on. Large interannual variability in the snowfall occurs, with the wettest year on record having ten times as much snow as the driest year. Along with spatial variability in snow accumulation, the rate of snowmelt varies with location and with deposition of light-absorbing particles in the snowpack that cause more absorption of solar radiation. Study of the mountain snowpack is difficult because of the lack of regular measurements at the high elevations. As climate changes, there is an increasing need for improved methods of observation, including remote sensing, to drive a more mechanistic understanding of snow processes.

Key Words. snow, snowmelt, snow accumulation, spatial variability, solar radiation, remote sensing

SNOW AS A SOURCE OF WATER FOR
LAKES AND WATERSHEDS

The Sierra Nevada accumulates substantial amounts of snow. At higher elevations, snow covers the ground for at least half the year; water is thus stored in its frozen form for many months, with little water released to the watershed until melt begins. Of the seasonal hydrologic changes that affect soils, streams, and lakes in the Sierra, the most significant are the winter accumulation of snow, the spring and summer melt of the snowpack, and the infrequent precipitation in summer and early autumn. This wet-dry seasonal difference, combined with rugged topography, steep gradients in temperature

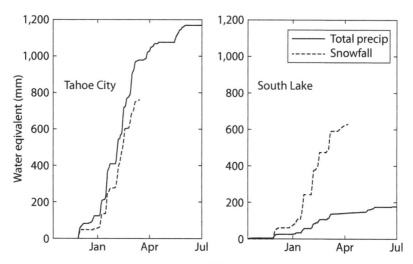

FIGURE 8. Total precipitation and snowfall at two locations in the Sierra Nevada, Tahoe City (39.171°N, 120.160°W, 2,057 m elevation) and South Lake (37.176°N, 118.560°W, 2,926 m elevation) in WY 2019. At the lower elevation, Tahoe City, total precipitation slightly exceeds snowfall until late March, the difference between total precipitation and snowfall attributed to rain. Later in the season, all the precipitation occurs as rain, the snowpack peaking around the beginning of April. At the higher elevation, South Lake, the snowfall appears to exceed the precipitation; this artifact likely results from almost all the precipitation coming in the form of snow and the rain gauge that measures total precipitation failing to catch most of the snow, a known problem in hydrology of cold regions (Yang et al. 2000).

and precipitation with elevation, and high interannual variability, makes hydrologic processes and variations in the high-elevation Sierra Nevada significantly different from those at lower elevations. Because of its persistence through spring and often into the summer, the snowpack provides soil moisture for vegetation during the part of the year when evapotranspiration increases. At the higher elevations, snow protects vegetation from damage by winter winds. Streams and lakes receive a large influx of water and solutes from spring runoff. At lower elevations, much of the precipitation occurs as rain (figure 8). Rainfall rates can be higher than snowmelt rates because the derivative of saturation vapor pressure versus temperature is greater at higher temperatures, so orographic lifting of warm, saturated air produces greater precipitation rates than the rate of snowmelt even under the most favorable conditions Therefore, the river basins at lower elevations, for example, in the northern Sierra, are more prone to winter floods, whereas in the high-elevation basins of the southern Sierra winter rain is rare.

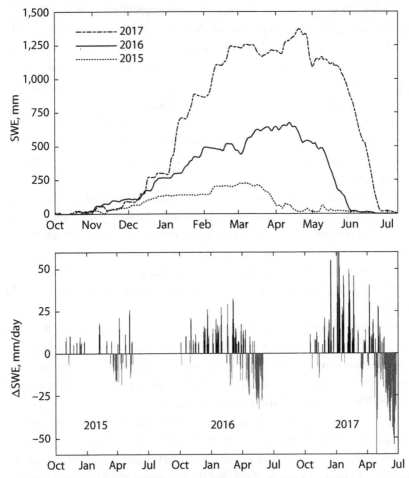

FIGURE 9. Data for 2015–17 from a snow pillow at Dana Meadows (identifier DAN in the California Department of Water Resources data system, http://cdec.water.ca.gov /snow/current/snow/index.html) at 2,987 m elevation in the Tuolumne River drainage. The year 2015 was the driest on record, 2017 was the wettest year since 1983, and 2016 was a near-average year. The graph on the top shows the accumulated snow as millimeters of snow water equivalent (SWE). The graph on the bottom shows the daily differences of SWE, with the positive direction representing daily snow accumulation and the negative representing daily melt. At high elevations, little melt occurs before April. The melt rate is greater in the wetter years because more of the melt occurs when the daylight hours are longer, the sun elevation is higher, and the air temperatures are warmer.

Snowfall and snowmelt in the Sierra follows a typical progression (figure 9). These data are from a snow pillow, which measures the pressure (and thus the weight, i.e., the mass times the gravitational force) of the overlying snowpack. The snowpack's mass can be converted to the snow water equivalent (SWE), expressed either as a mass per unit area (kg m^{-2}) or as an equivalent depth of water in millimeters by dividing by the density of water, ρ_w, so 1 kg m^{-2} is equivalent to 1 mm depth. Total snow water equivalent SWE = D × ρ_S, the snowpack's depth D times the average snow density ρ_S. Density of newly fallen snow can be less than 50 kg m^{-3}; density of melting and refrozen snow typically exceeds 400 kg m^{-3}.

Distinct from the snow water equivalent is the liquid water content. As snow melts at the surface and water drains into the snowpack, capillary tension holds the liquid water in the spaces between the snow grains. Snow's coarse grains are similar in size to the particles in sandy soil, so the field capacity, the maximum liquid water content that the snow will hold against the force of gravity, is about 10% unless frozen impermeable layers in the snowpack impede drainage.

Liquid water at the soil surface leads to runoff, a key component of the water balance of a watershed, which describes how water is partitioned among runoff, evapotranspiration, and fluxes into and out of the soil water and groundwater, as described in chapter 4. An important difference between rainfall and snowmelt is that when precipitation occurs as rain, the water is rapidly available as runoff or to recharge groundwater, whereas when it occurs as snow, weeks or months may elapse before melt occurs. For water resources for human uses, this delay enables a forecast of runoff well before the runoff occurs. For ecosystems, this delay in the timing of water availability strongly affects watershed biogeochemistry and the limnology of Sierra waters, as described in chapters 5 and 6.

MEASUREMENTS OF SNOW

Snow Depth, Density, and Water Equivalent

In response to the importance of snow as a water resource, water management agencies in the river basins draining the Sierra Nevada began measurements of depth and SWE in 1910. These measurements are made for operational forecasting rather than scientific research, though the scientific analysis of data that were collected for other reasons is common in studies of hydrology and climate (National Research Council 1991).

Falling snow is notoriously hard to catch in a precipitation gauge (Yang et al. 2005). Alternatively, since snow remains in place for a while after it falls, its accumulation is more accurately assessed by measuring its depth and water equivalent on the ground. A snowboard can be used to measure snowfall from an individual storm. It is placed on the snow surface before the storm; then afterward the depth and density of the newly fallen snow are measured. Snow accumulation and settling over a season are most accurately measured in a snowpit; these require considerable effort to dig and are usually used when information is needed about the density and temperature profile, presence of weak layers, presence of compounds other than H_2O, or other stratigraphy (Greene et al. 2016).

More commonly used measurements enable routine data collection at multiple sites. As of 2020, at 259 snow courses in the Sierra Nevada, regular manual measurements are made by plunging a tube through the snow to the underlying soil surface, extracting a core of snow whose length is the depth of the snowpack, and then weighing the tube and thereby calculating the snow water equivalent. James Church, a professor of classics at the University of Nevada, Reno, invented this technique in the early twentieth century (Church 1933). The first snow course in the Sierra Nevada was established on Mount Rose in 1910, and such a tube is referred to as a Mount Rose snow sampler. At most of the courses, snow surveyors traveling on skis (usually) or over-snow vehicles begin measurements around February 1 and continue near the first of each month through April; in years with deep snowpacks, measurements continue through May and occasionally June (Armstrong 2014). Of the 259 courses, 197 have seventy or more years of data. Snow pillows measure the pressure of the overlying snowpack at the soil surface, and the data are transferred by telemetry. Of the 130 pillows operating through 2020, 124 have thirty or more years of data.

A problematic attribute of the snow courses and pillows is that they do not sample the topographic variety. For logistical reasons, snow pillows and other remote meteorological sites in the mountains lie on nearly flat terrain, so they may poorly represent snow accumulation and melt rates on nearby slopes (Meromy et al. 2013). For instance, in midwinter the snow pillows on the valley floor and all slopes nearby are covered with snow, but later in the spring the valley floor still has snow while slopes facing the sun are bare. In addition to snow pillows not covering the slopes and other aspects of topography, the highest ones in many basins are 1,000 m or more below the highest elevations, with the result that significant snow remains after it has melted from all the pillows. This snow provides runoff to high-elevation lakes, soil moisture, and groundwater.

Remote Sensing of Snow-Covered Area, Albedo, and SWE

The extent of the snow cover, perhaps its most important characteristic, can be measured by satellite. The high reflectivity of snow in the visible part of the solar spectrum enables satellite measurements of snow-covered area (Dozier and Painter 2004). In slightly longer wavelengths in the near-infrared, ice itself is more opaque, so snow's reflectivity depends inversely on grain size. This sensitivity to the size of the scattering particle also allows discrimination of clouds from snow. Examples of multispectral sensors that remotely sense snow cover, and whose data are freely available, are the Thematic Mapper (TM and ETM+) and Operational Line Imager (OLI) on the Landsat series of satellites (Dozier 1989; National Research Council 2013), the European Space Agency's Sentinel-2a and Sentinel-2b (Hollstein et al. 2016), the Moderate-Resolution Imaging Spectroradiometer (MODIS) on NASA's Earth Observing System Terra and Aqua satellites (Hall et al. 2002; Painter et al. 2009), and the Visible Infrared Imaging Radiometer Suite (VIIRS) on Suomi NPP (Key et al. 2013).

In measuring snow cover from satellite, one faces choices in temporal, spatial, spectral, and radiometric resolution; spatial coverage; and dynamic range. For mapping mountain snow, the combination of the Landsat and Sentinel-2a,b sensors provide 10 m to 30 m spatial resolution at four-day intervals, whereas MODIS and VIIRS see the whole Earth at 500 m and 375 m resolution, respectively, twice daily, with half those observations at night. Because snow changes rapidly and frequently, the daily temporal frequency of MODIS or VIIRS enables characterizing the behavior of the snowpack. However, the coarser spatial resolution compared to Landsat and Sentinel means that MODIS or VIIRS will miss the topographic variability of snow in the mountains. A solution to this conundrum is to use a spectral mixing model to estimate the fraction of snow in each 500 m MODIS pixel (Painter et al. 2009) and validate those snow fractions with Landsat or Sentinel over a wide suite of sample images (Rittger, Kahl, and Dozier 2013). In the Sierra Nevada, Dozier et al. (2008) have documented the need to use a spectral mixing model. They show that one-third of the snow cover typically occurs in MODIS 500 m pixels whose snow cover is less than one-half.

The albedo of the snow (the ratio of reflected solar radiation to the incoming magnitude) determines how quickly the snow will melt. Integrated over the solar spectrum, snow albedo in the Sierra Nevada varies from about 0.4 to 0.95 (Bair, Davis, and Dozier 2018; Seidel et al. 2016), so the absorption varies from 5% to 60% of the incoming solar radiation. Snow is brighter than all other surfaces in the visible part of the spectrum but is darker at wavelengths beyond 1.0 μm. Ice itself is transparent in visible wavelengths; beyond the visible in the near-infrared wavelengths, however, ice is more absorptive (Warren and Brandt 2008). Therefore, snow is darker in the near-infrared and its reflectivity depends on grain size (Wiscombe and Warren 1980). Hence in addition to mapping snow cover from satellite, its spectral signature allows estimation of whether it is fine newly fallen snow or coarser snow that has undergone metamorphism or melt-freeze cycles (Dozier et al. 2009).

With current remote sensing technology, it is not possible to reliably measure SWE by satellite in mountainous terrain. Therefore, estimating the spatial distribution of snow depth and water equivalent in mountainous terrain, characterized by high elevation and spatially varying topography, is one of the most important unsolved problems in mountain hydrology. One such effort is the NASA/JPL Airborne Snow Observatory (ASO) that carries a lidar (light detecting and ranging) altimeter to measure snow depth by subtracting the elevation of bare ground, measured during summer, from the elevation of the top of the snowpack (Painter et al. 2016). Combining depth with snow density, measured at snow pillows and modeled elsewhere, yields SWE.

In a review of remote sensing in hydrology generally, Lettenmaier et al. (2015, 7333) argue that "among all areas of hydrologic remote sensing, snow (SWE in particular) is the one that is most in need of new strategic thinking from the hydrologic community."

Deterministic and Probabilistic Reconstruction of Snow Water Equivalent

Data from snow pillows and courses can be used as statistical indices of the total volume of snow in the drainage basin for the purposes of estimating runoff. However, calculating a water balance for a specific basin requires that the amount of snow be assessed for that basin. In a small basin such as the Emerald Lake watershed, such assessments are possible with intensive ground measurement (see chapter 4), but those methods are impractical for frequent or spatially extensive coverage.

Satellite-based measurements of snow cover indicate when the snow has been depleted; hence it is possible to reconstruct the time series of SWE during the melt season. Originally implemented by Martinec and Rango (1981), the method works as follows: Starting with the date of the disappearance of snow, calculated snowmelt is summed backward daily to estimate the accumulated SWE for each day back to the last significant snowfall. SWE on day N is calculated by starting with SWE at the end of day 0 and summing the melt M for each day:

$$SWE_N = SWE_0 - \sum_{j=1}^{N} M_j$$

Then, if the value of N for which $SWE_N = 0$ is known, one can solve for SWE_0. The method has two limitations. First, reconstruction is possible only back to the last significant snowfall, which in the Sierra Nevada is usually adequate if based on an April or May snow survey; and second, the reconstruction cannot be used to forecast spring runoff.

The technique has been validated in the Sierra Nevada (Cline, Bales, and Dozier 1998) and applied to large basins at multiple scales (Molotch and Margulis 2008). Recent work comparing forward-based modeling and reconstruction suggests that in areas with high precipitation uncertainty, reconstruction provides the best approach for SWE estimation (Raleigh and Lundquist 2012). In the Sierra Nevada, the snow depletion from reconstruction better matches observed streamflow than does interpolation from snow pillow measurements or from a numerical model assimilated with surface data (Dozier 2011; Rice et al. 2011). The more accurate results employ energy-balance snowmelt models (Homan et al. 2011; Rittger, Kahl, and Dozier 2011), along with integration with the fractional snow cover, to recover each pixel's SWE (Bair et al. 2016; Margulis et al. 2019) and its heterogeneity (Kahl et al. 2014).

SPATIAL AND TEMPORAL DISTRIBUTION
OF SIERRA NEVADA SNOW

Measurements of snow on the ground and from remote sensing have been used to examine several characteristics that are relevant to our examination of high-elevation lakes and the hydrology of their watersheds, including (1) interannual variability of the amount of snow and the timing of its melt,

(2) relationships between snowfall and large-scale climate indicators such as El Niño and the Pacific Decadal Oscillation, and (3) possible existence of secular trends related to human-induced climate change. For example, most snow pillows in the western United States show that there has been a decline in the April 1 snowpack since 1950 and that the spring snowmelt pulse in streamflow occurs earlier (see chapter 7 for further discussion of these topics).

Historical Variation from Snow Pillows and Snow Courses

A characteristic of precipitation generally is its "long-tailed" distribution. At the low end, the depth and water equivalent of snow cannot be negative, whereas at the high end, the wettest years can experience three times more snow accumulation than the mean value. Figure 10 illustrates the annual progress of snow accumulation for the wettest and driest years in the period since the early 1980s when the spatial density of snow pillows was large enough to provide a daily picture of most of the range. Figure 11 shows the progression of SWE for the years 1983 through 2019, based on scattered interpolation in three dimensions (Northing, Easting, Elevation). In the wettest year, 1983, the snow continued to accumulate in the spring and the peak accumulation occurred in early May. The rate of meltwater production in the wettest years is higher than in the average or dry years because the melt occurs in June and July when the sun elevation is high, the daylight periods are long, and the air temperatures are warm. In the driest years, snow accumulation ceases earlier, melt starts in March, and the rate of melt is slower because less solar energy is available. For these reasons, Musselman et al. (2017) assert that in a warming climate, the rate of snowmelt would slow because it would begin earlier in the spring.

The long period of record of the snow course data enables examination of patterns of precipitation and possible trends over the period of record. Aguado (1990) concluded that in most years the Sierra Nevada behaves nearly uniformly; wetter than average years are wet throughout the range, and the same is true for years that are drier than average. Some years, however, show a pattern in which high- and low-elevation sites have opposite departures from average. Over the entire period of record he analyzed, this pattern accounted for a small percentage of the total variance, although in some years it was conspicuous.

Possible trends in April 1 SWE as a function of latitude and elevation are complicated (figure 12). Snow courses that show a statistically significant

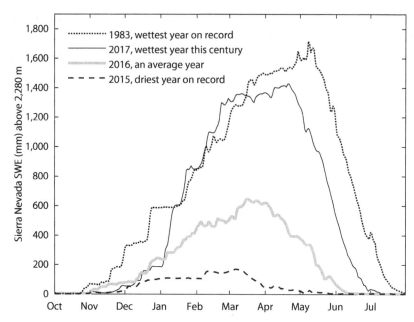

FIGURE 10. Progression of SWE through the year as a spatially weighted average over all high-elevation (2,280 m or higher) snow pillows in the Sierra Nevada for the wettest year on record (1982–83), the wettest year this century (2016–17), the driest year on record (2014–15), and an average year (2015–16).

declining trend over the past 50+ years are at the lower elevations. All analyses based on the long-term snow course record should be treated with caution, for several reasons:

1) The April 1 measurement may be an unreliable surrogate for the maximum snowpack or the total winter accumulation, and the literature disagrees about the likelihood that the choice of a single date could introduce error. Bohr and Aguado (2001) show that the April 1 SWE underestimates the peak by 12% in the Rocky Mountains but provides a more accurate estimate, 4%, of the total season accumulation. Kapnick and Hall (2010) employ an interpolation method to estimate the date of maximum SWE and show that since 1930 there has been an overall trend toward earlier snowpack peak timing by 0.6 day per decade. The trend toward earlier timing occurs at nearly all individual stations, even those showing an increase in April 1 SWE. Montoya, Dozier, and Meiring (2014) show that the use of April 1 introduces the largest error at the two ends of the Sierra Nevada; peak snow generally occurs in

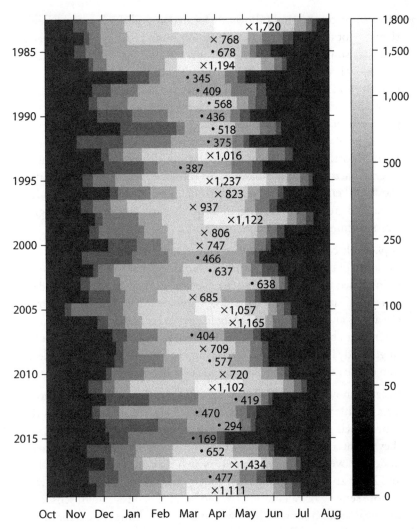

FIGURE 11. Summary of spatially distributed SWE, in mm, based on spatial interpolation from snow pillows. The × and · symbols indicate the date of maximum accumulation each year, × for years wetter than the median and · for the drier years.

early May in the high elevations in the south and in early March at low elevations in the north. Moreover, later peaks are associated with El Niño years and earlier peaks with La Niña, especially when the Pacific North American (PNA) index is colder than normal. However, the magnitude of bias (the difference between the maximum accumulation and that on April 1) is greatest when the peaks are earlier.

2) Most snow courses and pillows are located at middle elevations; hence they do not represent trends at the lower boundaries of the snowpack, and they sometimes show that all snow has disappeared when considerable snow is visible higher. Dettinger (2014) expresses the concern that the measurement network has missed a plausible decrease in the orographic enhancement of precipitation (which is determined by the horizontal wind speed, atmospheric water vapor concentration, and slope of the mountains in the direction facing the wind) in the Pacific Northwest, owing to a slowdown in the westerly winds that Luce, Abatzoglou, and Holden (2013) observed. Climate projections show weakened lower-troposphere zonal flow across the region under enhanced greenhouse forcing, highlighting an additional hydrologic stressor from climate change.

3) A plausible consequence of a warming climate is an increase in the elevation of the rain-snow transition, a change that would likely lead to increased winter runoff and decreased summer streamflow (Field et al. 1999). Currently, the snow pillow network's lack of coverage at low elevations and the inability of rain gauges to accurately measure snowfall mean that observations of such a trend are mostly anecdotal, except for a single study in southern Idaho that showed a rise of 300 m in the elevation of the rain-snow transition in the past fifty years (Marks et al. 2013).

4) Because snow courses involve transects of 100 m to 200 m, they have a reasonable chance of representing the area within about 1 km^2 of the course itself. Snow pillows, however, are from 1 m^2 to 8 m^2 in area, so they essentially represent a point measurement. Thus they can either under- or overrepresent the area in which they are located (Molotch and Bales 2005), hindering their use in validating measurements from remote sensing or calculations by models. Embedding small sensors in the areas around snow pillows would help determine the extent of this problem (Rice and Bales 2010).

Snow-Covered Area

Remotely sensed snow-covered area provides information about the spatial distribution of snow that is not available from snow pillows or courses. Late in the snowmelt period, the standard measurements would indicate that snowmelt as a source of water for soil moisture and runoff has ended for that year. Yet Landsat images of the Tuolumne and Merced River basins in the Sierra Nevada show that in 2011 significant snow remained at elevations

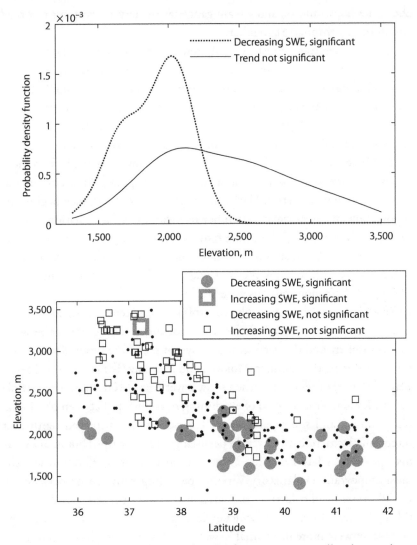

FIGURE 12. Long-term trends in April 1 SWE from snow courses, all with more than 50 years of records and extending into 2019. The scatterplot shows that the snow courses with statistically significant decreasing trends (at the *p=0.10* level) lie at low elevation, below the 2,280 m threshold corresponding to high-elevation lakes. The only course with significantly increasing snow lies at 3,300 m elevation. The histograms in the upper graphs show that the distribution of elevations with trends differs from the elevation distribution of courses in general.

above the snow pillows, and stream gauges at the lower elevations recorded runoff in July through September.

Snow is heterogeneously distributed in the montane environment. Orographic processes cause snowfall to increase with elevation, but in the Sierra Nevada snowfall tails off above about 3,000 m as water vapor in the air mass condenses. The magnitude of the orographic effect in the Sierra Nevada varies year to year depending on the direction of the storm track and the temperature and water vapor concentration of that storm (Lundquist et al. 2010). Once on the ground, wind redistributes snow, scouring it from more exposed locations and depositing it in protected spots (Elder, Dozier, and Michaelson 1991; Winstral, Marks, and Gurney 2013). Once melt begins, the topography and trees cause variability in incoming solar radiation (Dozier 1980; Hardy et al. 2004), longwave radiation (Flerchinger et al. 2009; Marks and Dozier 1979), and sensible and latent heat exchange (Marks et al. 2008; Reba et al. 2009), so snow melts at widely different rates over the spatial domain of these mountainous watersheds.

The Sierra Nevada topography controls the persistence of the snow cover, and the persistence in turn influences soil moisture, energy exchange between the atmosphere and the surface, the ecosystems, and the input of water into the lakes during the summer. Molotch and Meromy (2014) examined seven years (2001–7) of remotely sensed snow cover from MODIS over the whole Sierra Nevada, with the goal of examining relationships between physiography, climate, and snow persistence. The persistence displayed significant interannual variability, on average persisting into early April but in dry years disappearing by late February. In most basins (10 of 13) elevation was the most important explanatory variable regarding snow cover persistence. Locally, pooling of cold air creates areas where snow persists, possibly establishing future refugia for snow-dependent species (Curtis et al. 2014). Slope and aspect were more influential in watersheds with a north–south orientation. Such results imply that orientation may have more influence as climate warms, because the solar geometry creates more heterogeneity with the topography in years when snow melts earlier (because during the daylight period, the sun goes through a smaller range of azimuths).

Reconstructed Snow Water Equivalent

Using Landsat imagery to map snow cover and a precipitation model to estimate snowfall, Margulis et al. (2016) reconstructed SWE for the Sierra

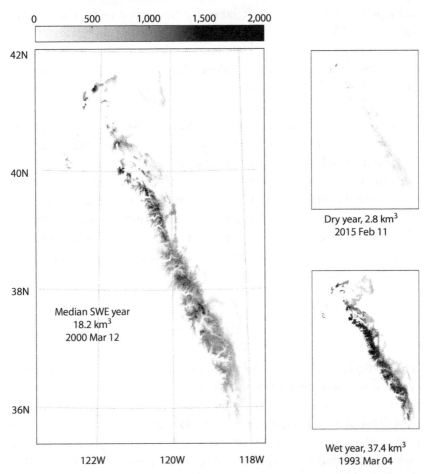

Snow Water Equivalent (SWE), mm

Dry year, 2.8 km³
2015 Feb 11

Median SWE year
18.2 km³
2000 Mar 12

Wet year, 37.4 km³
1993 Mar 04

FIGURE 13. Variability in SWE for the period 1985–2016, based on data from Margulis et al. (2016a). Their analyses combined Landsat imagery of snow-covered area with precipitation modeling. The images coincide with the maximum values in each year. Shown are the year nearest the median value (2000) and values for the driest year (2015) and the year with the greatest snow accumulation (1993). The gray bar above the median image shows the SWE in mm.

Nevada for water years 1985 through 2015; subsequently, they extended the record through 2016. Usually, drier years reach their maximum accumulation earlier than the wetter years, but numerous exceptions occurred. As one would expect, the snow persists longer in wetter years.

Characterizing the interannual and seasonal variability of snow requires information about its spatial distribution, Figure 13 shows snow water equivalent for a year near the median value of the annual maximum SWE and representative dry and wet values for the period 1985–2016 over the whole Sierra Nevada, as reconstructed from the depletion history of each pixel. For each day, the melt at each pixel is calculated as the product of the one-dimensional melt rate $M_j(P)$ multiplied by the fractional snow cover, i.e., $Mj = M_j(P)F_{snow}$. The melt rate is calculated from estimates of solar radiation, longwave radiation, air temperature, and humidity by the North American Land Data Assimilation System (Cosgrove et al. 2003), downscaled from its 1/8° resolution to the 500 m resolution of the MODIS data using topography from the Shuttle Radar Topography Mission (Farr et al. 2007).

In the driest year, 2015, snow water equivalent was below the median throughout the range. At the higher elevations in the middle of the range, there was less variability, as those areas are nearly completely snow covered in all years. In the wetter years, there was more snow than normal at the lower elevations, because the higher elevations have widespread snow, either deep or shallow.

MODELING SNOWMELT

A physically based model of snowmelt, requiring no site-specific calibration, considers the mechanisms by which the snowpack can gain or lose energy. In the Sierra Nevada, and indeed in almost all environments worldwide where snow dominates the hydrologic cycle, net solar radiation drives the snowpack energy balance. Net solar radiation, positive during the daytime and zero at night, is the product of the incoming solar radiation and the fraction absorbed, which is 1.0 minus the snow albedo.

The other component of the radiation balance is the net longwave radiation, balancing that emitted downward by the atmosphere and emitted upward by the snow surface. The net all-wave radiation is almost always negative at night and usually positive during the day. Integrated over the day, the net radiation is typically negative during the winter when the daylight period is short and becomes positive in the spring, thereby driving the spring pulse in snowmelt in the Sierra Nevada (Marks and Dozier 1992).

The other major energy balance components, sensible and latent heat exchanges and heat conduction downward into the snowpack or upward out of it depending on the direction of the temperature gradient, are smaller but help determine whether the snow is at 0°C, the melting temperature of ice.

In most conditions during spring and early summer in the high-elevation Sierra Nevada, sensible heat exchange goes toward the snowpack (air temperature is warmer than the snow), but latent heat exchange goes in the opposite direction (vapor pressure in the air is less than the saturation vapor pressure at the snow surface temperature).

In the Sierra Nevada, the snow study site on Mammoth Mountain (Bair et al. 2015), maintained by the University of California, Santa Barbara (UCSB), and the U.S. Army's Cold Regions Research and Engineering Laboratory (CRREL), is one of only half a dozen sites in the United States that measure the snowpack energy balance with all instruments carefully calibrated and maintained. From a data set from 2011 through 2017 expressly tuned to drive and validate snowpack energy balance models (Bair, Davis, and Dozier 2018), we note the following observations of the energy balance that characterize the high-elevation Sierra Nevada:

- Net daily average solar radiation with snow on the ground averaged about 53 W m^{-2}, ranging from about 15 to 170 W m^{-2}; net daily longwave radiation averaged about −46 W m^{-2}, ranging from about −70 to −30 W m^{-2} with occasional positive values on cloudy days. As the solar elevations and the length of daylight increase in the spring, the net all-wave radiation (the sum of the net solar and the net longwave) becomes positive, and spring melt is sustained (figure 14).

- Snow albedo ranged from 0.38 to 0.95, with a modal value of 0.8, the lowest values corresponding to periods when dust from local sources contaminated the snow.

- The vapor pressure gradient was toward the snow surface 7% of the time during the years 2011 to 2017, so latent heat mostly flowed away from the snowpack, indicating mostly evaporation instead of condensation.

- Snow density ranged from 50 to 680 kg m^{-3}. The median value was 375 kg m^{-3}, with an interquartile range from 300 to 435 kg m^{-3}.

FUTURE DIRECTIONS

Remote Sensing of Snow Properties

The network of snow pillows does not cover the highest elevations, nor does it cover the lower elevations, because the observations supported forecasts of the spring and summer snowmelt runoff and were not designed to address

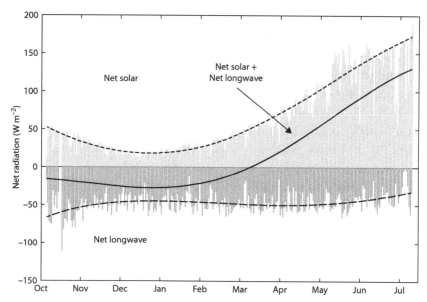

FIGURE 14. Components of daily net radiation balance at Mammoth Mountain at 2,940 m, representative of the high-elevation Sierra Nevada, showing median daily progression of net solar and net longwave radiation over WYs 2011–17 during the months while snow is on the ground (data from Bair, Davis, and Dozier 2018). The lines show smoothed values. Typically, sometime in March or April net all-wave radiation (the sum of net solar and net longwave) becomes positive, triggering sustained spring snowmelt (Marks and Dozier 1992).

scientific questions or to identify trends. Moreover, all pillows are on flat ground. Therefore, there are no operational measurements of snow on slopes, so the SWE values presented in this chapter provide indices of the amount of snow in a watershed, but they do not provide the volume of water stored in a watershed's snowpack. Although airborne measurements of snow depth using lidar are now being done with the Airborne Snow Observatory (Painter et al. 2016), and SWE is calculated by multiplying depth by estimated density, the technology is not easily transferable to satellites. Monitoring trends in the hydrology of high-elevation lakes in the Sierra Nevada would benefit from improvements in the science and technology of remotely sensing snow water equivalent (Painter et al. 2016).

Estimating the Radiation Balance

Snowmelt rates and dates of disappearance depend mainly on the net solar and longwave radiation, which in turn depends on the fractional cover of snow and ice and their albedo and surface temperature. Regardless of how snow depth or water equivalent is measured, the retrievals must be accompanied by an energy balance model that simulates characteristics of the snowpack, and any model chosen will require local, relevant energy balance information.

The information needed at the scale of mountain topography includes atmospheric properties to calculate the downwelling solar and longwave radiation, the spatial and temporal variability of albedo of the snow in areas where vegetation and topography also affect any remotely retrieved signal, and the temperature of the snow separately from the temperature of the other components of the scene.

Measuring and Modeling Snowfall

To complement measurements of snow that already has fallen to the surface, integration of the surface hydrology with the meteorology would be fruitful. Automatically measuring falling snow at a single site is problematic, especially during windy conditions, and like the snow courses and snow pillows, precipitation gauges would not cover the landscape even if they were reliable. Therefore, remotely sensing precipitation in its frozen state and accurately modeling storms in the mountain environment would be a significant achievement.

Models of snow-dominated precipitation in the mountains produce problematic, generally unvalidated results. The widely used SNODAS model (Barrett 2003) is not described in useful detail in any peer-reviewed publication or, as far as we know, even in the gray literature. Comparisons of its output with independently measured spatially distributed snow accumulation in the mountain environment show significant over- and underestimates (Bair et al. 2016; Clow et al. 2012; Hedrick et al. 2015). The model's operational documentation does not specify which surface measurements have been assimilated, so operational surface measurements from snow pillows cannot be compared as a validation source.

Measurement of snow water equivalent in mountain regions therefore remains elusive. In accessible ranges like the Sierra Nevada, periodic measurements of snow depth from airborne lidar altimetry provide a feasible method to measure the spatial distribution of snow throughout the winter and spring (Kostadinov et al. 2019; Painter et al. 2016). To interpret the magnitude of the snow resource between airborne acquisitions, Margulis et al. (2019)

consider a data assimilation approach to generate space-time continuous estimates of snow water equivalent from occasional snow depth measurements from aircraft. Similarly, Hedrick et al. (2018) model the rate of snowmelt between flights to calculate daily values of the distribution of snow. These research efforts integrate measurements with models to take advantage of the confidence and accuracy of the data with the ability of models to overcome constraints of spatial and temporal frequency.

FOUR

Watershed Hydrology

Abstract. Hydrologic characteristics of high-elevation watersheds are strongly related to their biogeochemistry, ecology, and responses to environmental changes. Seasonal and annual variability in the inputs and outputs of water in the Sierra Nevada reflects variations in climatic conditions. Inputs (snow and rain) and losses (stream flow, evaporation) of water for seven headwater watersheds and the upper Marble Fork of the Kaweah River (also referred to as the Tokopah basin), with a focus on the Emerald Lake watershed, are presented. Methods of measurement and uncertainties are briefly described. Most precipitation occurs as snow during the months of December through April. The majority of outflow occurs via streams during the snowmelt period, with a smaller annual percentage lost to evaporation. Measurements of precipitation, outflow discharge, and estimates of evaporation allow calculations of annual water budgets for the Emerald Lake watershed for three decades, for the Tokopah basin for two decades, and for six additional watersheds for four or five years. Snow was the dominant input of water to Emerald Lake and the Tokopah basin in all years except for 2015, when rainfall exceeded snowfall in what is the smallest snowpack on record in the Sierra Nevada.

Key Words. precipitation, rain, snow, stream discharge, evaporation, water balance

HYDROLOGIC CHARACTERISTICS AND BUDGETS of high-elevation watersheds are crucial to our understanding of their biogeochemistry, ecology, and responses to environmental changes. Seasonal and annual variability in the inputs and outputs of water in the Sierra Nevada reflect climatic conditions and determine flows that reach lower elevations (Kattelmann 1996). Fluxes of solutes and biological and geochemical aspects of watersheds are related to hydrologic processes. Kattelmann and Elder (1991) were the first to report measurements and calculations of all the components in the water balance for a high-elevation basin in the Sierra Nevada. Their approaches and analyses formed the foundation for the long-term studies at Emerald Lake and throughout the Sierra Nevada.

The annual water budget for a watershed consists of the hydrologic fluxes into and out of the watershed:

$$\Delta S = P - (Q + E) \pm G \pm \varepsilon$$

where:
   ΔS is the change in water storage;
   P is precipitation including rain and snow;
   Q is stream discharge, out from the watershed;
   E is evaporation, sublimation, and transpiration;
   G is groundwater exchange; and
   ε represents quantities not included or measurement error.

In the case of our analyses of water balances of Sierra watersheds, groundwater fluxes to the watersheds were not included since groundwater fluxes from the watersheds were usually quite small. Hence we simplify the equation:

$$\Delta S = P - (Q + E) \pm \varepsilon$$

In this chapter we present major inputs and losses of water for seven headwater watersheds and the upper Marble Fork of the Kaweah River (also referred as the Tokopah basin) with a focus on the Emerald Lake watershed. The characteristics of these watersheds are provided in chapter 2. A water year (WY) in California is defined as the period from October 1 of the previous calendar year to September 30 of the water year. In the Sierra Nevada, most watersheds include lakes; those that we consider in this book all do. As discussed in chapter 3, precipitation from November through March or April typically falls as snow, while precipitation from April or May through October usually falls as rain.

METHODS AND UNCERTAINTIES

A thorough treatment of the methods for and uncertainties in determination of water budgets is provided in Melack et al. (1998), from which we derived a portion of the following material. Rainfall was measured with tipping bucket rain gauges usually located within each watershed. The number of rain gauges ranged from ten in the Tokopah basin to one in the Spuller Lake basin. Based on Winter (1981) and Kattelmann and Elder (1991), we estimate that the uncertainty for rain measured by the gauges was ± 5%. In early autumn or late spring some gauges were not deployed, and some precipitation fell as nonaccumulating snow or occasionally as rain. We extrapolated rainfall in cases when rain gauges were inoperable or not installed using measurements made at nearby weather stations, including the Lodgepole Ranger Station.

Accumulated snow was measured in surveys conducted at maximum snow-pack accumulation in late March or early April; this approach has the added advantage that snowpack evaporation and sublimation are incorporated and do not require separate calculation. Basinwide winter SWE was derived from snow depth surveys representative of the basin's topography and measurements of snow density. Using wedge-shaped cutters and portable electronic balances, snow densities were determined in snowpits dug to the ground. During the 1980s and 1990s, snow-covered areas in each watershed were obtained from aerial photographs. Subsequently, snow-covered area was based on satellite imagery (Perrot et al. 2014; Margulis et al. 2016a). Estimates of SWE for each year are based on many spatially distributed measurements of snowpack depth and several of density with estimated uncertainty of 10% to 15%. Adding SWE to the amount of rain yielded the total input of water for each catchment. During WYs with below average precipitation we estimate total uncertainty of 18%; and in WYs with above average precipitation, uncertainty of 11%.

Gauging discharge in Sierra streams presents challenges because streams are buried under snow for many months of the year, the channels are steep and boulder-strewn with turbulent flows, and most watersheds are located in federally protected wilderness areas that have restrictions on the construction of control structures. The gauging of water levels was done with pressure transducers installed in well-defined, natural cross sections augmented by staff gauges. In the case of the Emerald Lake basin after 1990 and at Spuller and Topaz Lake basins for their entire record, V-notch weirs were used. Discharge measurements in channels without weirs were determined from velocity profiles and tracer-dilution methods using either slug or constant injections (Kilpatrick and Cobb 1985; Herschy 1978). Stream discharges at the sites with V-notch weirs have uncertainties of ± 5%. Standard errors for discharge determined with slug injection was estimated to be ± 20%, while those based on constant injection were ± 7%.

Evaporation is the conversion of liquid water to vapor and its subsequent loss to the atmosphere, sublimation is the transformation from ice to vapor, and transpiration is the process by which plants take up soil moisture and release water vapor. We refer to all these processes as evaporation, or as evapotranspiration in figures. Evaporation in seasonally snow-covered mountain catchments is difficult to determine as there is considerable variation in reported magnitude, and all available methods have limitations, as discussed by Leydecker and Melack (1999). For WYs 1985, 1987, and 1990–94, catchment evaporation from snow was calculated with the mean-profile method using multiyear meteorological data from locations in or near each of the watersheds, as described by Leydecker and Melack (1999) and Marks and Dozier (1992). Annual evaporation was estimated by applying a model developed by Morton (1983) called complementary relationship areal evaporation (CRAE), as described by Leydecker and Melack (2000). For all other years, annual evaporation was estimated from a regression between snow water equivalence measured in the April snow survey and the residual of the measured water balance (rain + snow − outflow = residual). We assumed that the majority of the residual is accounted for by evaporation.

PRECIPITATION

Most precipitation in high elevation catchments of the Sierra Nevada occurs during the months of December through April. From 75% to over 95% of annual water deposition is snow (figures 15 and 16). A notable exception to this pattern occurred during the severe drought of 2015, when for the first time in our over 30-year record rainfall exceeded snowfall. The summer season can have frequent afternoon thunderstorms or can be almost rainless. Large spring and autumn storms occur in some years, but often these periods have little precipitation. Annual variability in precipitation is large. Over the 32-year record at Emerald Lake (1983–2016), annual nonwinter precipitation (mostly falling as rain) varied from 26 mm to 597 mm and winter snow deposition ranged from 139 mm to 3,177 mm.

Elder, Dozier, and Michaelson (1991) measured snow depths in 1986 through 1988 throughout the Emerald Lake watershed, along with snow density at half a dozen snowpits, to estimate the spatial distribution of SWE. They evaluated a stratified sampling scheme by identifying and mapping zones of similar snow properties on the basis of topographic parameters that

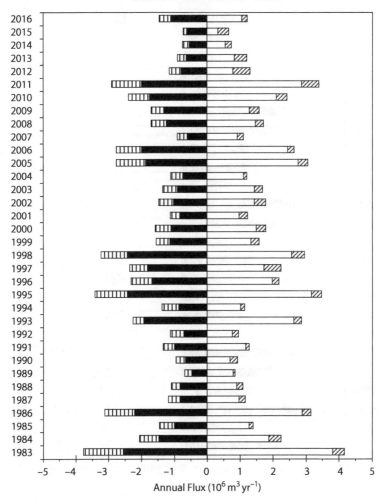

FIGURE 15. Annual inputs of snow and rain and outputs via stream discharge and evapotranspiration (ET) for the Emerald Lake watershed for WYs 1983–2016. Units: 10⁶ m³.

account for variations in both accumulation and ablation. Leydecker, Sickman, and Melack (2001) continued this work in the Tokopah basin from 1994 to 1999, based on up to 1,000 measurements throughout the basin. Using the concept of representative elementary area (REA; Wood et al. 1988), they determined an approximate REA of 50 ha. The consistent year-to-year relationship between mean basinwide snow depth and mean depths in subbasins, such as the Emerald catchment, suggests the possibility of snow

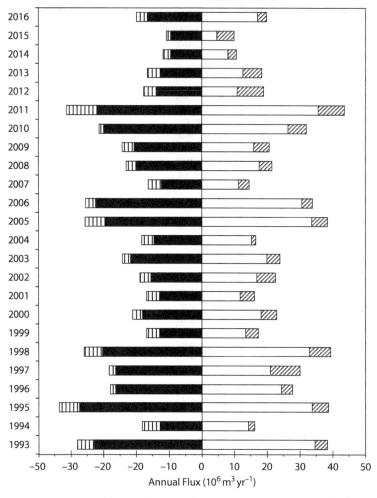

FIGURE 16. Annual inputs of snow and rain and outputs via stream discharge and evapotranspiration (ET) for the Tokopah basin for WYs 1993–2016. Units: 10^6 m³.

distribution modeling based on stratified sampling of subregions (Leydecker, Sickman, and Melack 2001). As a consequence, if a subbasin is adequately sampled, a good estimate for the whole Tokopah basin can be made.

Aside from deeper snow below avalanching slopes and shallower snow at lower elevations and on slopes subject to winter melt, individual snow depths in the Tokopah basin were randomly distributed, as indicated by semivariograms that usually had an autocorrelation range of < 30 m; that is, adjacent pixels with similar terrain and radiation inputs can have dissimilar snow depths (Leydecker, Sickman, and Melack 2001). Net winter solar radiation and other terrain-based parameters, for example, slope, aspect, and elevation, typically explained less than 15% of the variability of snow depth, whereas in the steeper Emerald basin, spring snow accumulation correlated with topographically driven incoming radiation. Applications of binary decision trees to estimate distributed SWE for some years in the Emerald and Tokopah basins have been successful (Molotch et al. 2005).

Though the records are shorter than at Emerald Lake in the other headwater watersheds, the multiyear data (1986–87 and 1990–94) on precipitation at the watershed scale for sites distributed throughout the Sierra provide valuable comparative information; chapter 2 provides descriptions of the additional watersheds. Spatial variations occurred in both the form and the amount of precipitation. SWE at the neighboring Pear Lake watershed ranged from 600 mm to 800 mm during the drought years of 1987 through 1992 and reached 2,000 mm during the winter of 1993. Snow deposition at Topaz Lake was similar to that at Emerald and Pear Lakes; in WYs 1987 through 1992 deposition ranged from about 500 mm to 700 mm, and in 1993 it was 1,200 mm. During WY 1993, rainfall at Emerald Lake was large compared to sites in the eastern and northern Sierra Nevada because precipitation from a series of spring storms fell as snow in the eastern and northern Sierra and rain in the southern Sierra. Rainfall at Lost Lake in the northern Sierra ranged from 108 mm in 1991 to less than 30 mm in 1993. Snow accumulation at Lost Lake was large compared to other sites and varied by more than a factor of 3. Because of the catchment's topography and the prevailing wind pattern, large amounts of snow blow into this basin from outside the watershed. Snow depths measured in 1993 exceeded 10 m in many spots, and the average depth was over 7 m. Elsewhere in the Sierra, similar snow depths are usually confined to areas below cliffs.

At Ruby Lake rainfall varied from 44 mm in 1993 to over 180 mm in 1987. Snow deposition ranged from about 550 mm to 650 mm over WYs 1987 through 1992 and in 1994, and in 1993 it was over 1,300 mm. Crystal Lake annual rainfall varied from 63 mm to over 120 mm during the period 1987 through 1993. Snow deposition ranged from 600 mm to 900 mm during the relatively dry period of 1987 through 1992 and was two to three times greater in 1993. Year-to-year variability in rainfall at Spuller Lake was large, ranging from nearly 190 mm in 1992 to less than 50 mm in 1993. The dry winters of 1990 through 1992 and 1994 had an SWE of about 600 mm to 900 mm;

during the wet winters of 1986 and 1993, SWE values were about 1,500 mm and 2,000 mm, respectively.

STREAM DISCHARGE

The majority of outflow from watersheds occurs during the snowmelt period that begins as early as late March in low snowfall years but may not start until early May when snowpacks are deep. Peak runoff occurs in early to mid-June when winter snowfall is light and during late June or early July in years with large snowpacks. In some catchments, peak discharge following a wet winter is much greater than in dry years (e.g., Lost and Topaz), while in others the range of peak flows is relatively small (e.g., Ruby and Crystal).

Snowmelt is often punctuated by periods of low discharge caused by spring snowstorms that lower the rate of snowmelt for several days. Following peak discharge, runoff recedes gradually during summer and autumn. Though groundwater storage and releases can contribute to year-round outflow, outflow streams, including the Marble Fork, can be dry by September. Winter runoff is typically low. Winter stream flow can be increased by displacement of lake water by snowfall and, to a lesser extent, from winter snowmelt from south-facing slopes, and rarely from avalanches striking the lakes. Hydrograph separations using isotopic and solute values performed for a year with annual precipitation of 2.4 m and for a year with annual precipitation of 1.3 m revealed that the contribution to discharge during the rising limb of the hydrograph from the previous water year was 10% to 20%, with 80% to 100% of snowmelt passing through soil and talus (Huth et al. 2004).

Discharge from the Emerald Lake watershed illustrates these general characteristics (figure 17). The outflow from the lake was dry after snowmelt runoff ceased in drought years, but modest flows persisted through the autumn when the previous snowpack was large. Autumn rain and snow sometimes kept the outlet flowing. During winter, flow was usually small and was composed primarily of displaced lake water from large snowstorms. Snowmelt began in late March or early April. Daily runoff gradually increased until it peaked in early June. Peak snowmelt discharge ranged from about 117,000 m^3 per month during drought years to over 900,000 m^3 per month in wet years. Peak daily discharge during snowmelt varied from about 12,000 m^3 d^{-1} to almost 45,000 m^3 d^{-1}; larger flows can occur during rainstorms. These variations reflect solar inputs as cloud cover and day lengths

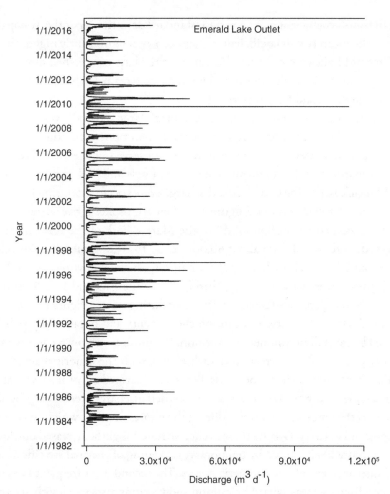

FIGURE 17. Stream discharge from the Emerald Lake watershed for WYs 1984–2016.

change and albedo is affected by occasional late-season snowfall, which increases albedo, and as dust and other light-absorbing particles accumulate on the surface, which decreases albedo.

Discharge patterns for the Tokopah basin are generally similar to those at Emerald Lake, except that spring runoff begins earlier and the river can dry earlier in the summer during years with small snowpacks. In years with deep snowpacks, peak monthly discharge was about 10^7 m^3 and as low as 5×10^6 m^3 in drought years. The runoff coefficient, defined as the dimensionless ratio of catchment runoff divided by precipitation, is often used to examine the rainfall-runoff response of catchments. The Emerald Lake watershed had a mean runoff coefficient of 0.70 ± 0.10 (standard deviation). For the Tokopah basin, the mean runoff coefficient, 0.85 ± 0.09, was significantly higher than at Emerald Lake (p < 0.001), indicating that the larger watershed more efficiently converts precipitation and snowmelt into stream discharge than does the smaller Emerald Lake watershed.

Abrupt increases in discharge attributable to large snowfall events or avalanches during the winter, rare warm winter conditions with rain-on-snow, and occasional intense rainstorms have been observed in the Sierra Nevada (Kattelmann 1990; Williams and Clow 1990; Engle and Melack 2001; Sadro and Melack 2012). The two highest discharge events occurred during January 1997 and October 2009 (see figure 17) when atmospheric river-driven discharge exceeded 1.2 million m^3 d^{-1} in the Marble Fork of the Kaweah (not graphed) and ranged from about 60,000 m^3 d^{-1} to 113,000 m^3 d^{-1} at Emerald Lake; outflow was above bank full discharge, with water flowing overland at both sites. Over the period 1985 through 2016 the Emerald Lake outflow hydrographs often reveal pulses of discharge during winter (figure 18), caused by large snowfall events, and rain-on-snow events and occasionally pulses caused by rainfall in summer and autumn. Examining outflow records for 32 years (1984–2016), we detected 54 such events, with the majority caused by avalanches following large snowfalls; flows were elevated only for a few hours to a day. Generally, the frequency of these pulsed discharge events was proportional to the amount of snowfall during the winter months, which peaks on average in February. During these events we noted slightly increased outflow temperature likely caused by temporary weakening of thermal stratification and displacement of hypolimnetic waters. The second most frequent type of winter discharge was caused by rain-on-snow storms associated with atmospheric rivers. These events were preceded by air temperatures above 0 °C and often generated much larger discharge pulses lasting several days (e.g., January 1997). The least common events were rainfall-driven spates and floods in summer and autumn that could produce very high flows with considerable biogeochemical effects on the lake (e.g., October 2009).

Williams and Clow (1990) calculated that a midwinter avalanche that struck Emerald Lake displaced about 70% of the unfrozen lake's volume, and Kattelman and Elder (1991) estimated peak discharge was between 10 and 20 m^3 s^{-1}, in comparison to peak snowmelt discharge the same year of about 1 m^3 s^{-1}. Sadro and Melack (2012) estimated that a rainstorm in October 2009 increased discharge from < 0.001 m^3 s^{-1} to over 3 m^3 s^{-1} and decreased the hydraulic residence time of the lake from 300 d to < 1 d.

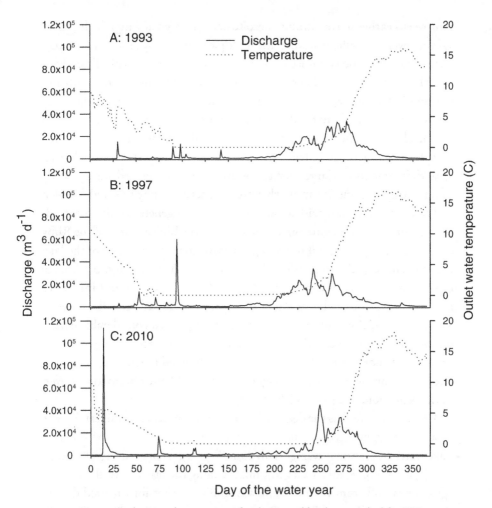

FIGURE 18. Stream discharge and temperature for the Emerald Lake watershed for WYs 1993, 1997, and 2010 illustrating specific features of hydrographs explained in the text.

The relative timing of snowmelt discharge (defined as the point where areal runoff consistently exceeds 1 mm day^{-1}) varied consistently in the watersheds studied. Catchments in the western Sierra Nevada (Emerald, Pear, and Topaz) had snowmelt begin three to four weeks earlier than catchments in the eastern Sierra Nevada (Spuller and Crystal) and up to five weeks earlier than at Ruby Lake (figure 19). Owing to their similar topography and aspect, snowmelt patterns were similar at Pear and Emerald Lakes; snowmelt at Topaz Lake produced similar peak discharge rates, but melt ended two to four weeks earlier in the summer despite being located at a higher elevation. Of the seven watersheds studied for multiple years, Ruby and Spuller Lakes were notable for the late initiation of snowmelt in spring (typically early to late May) and the long tail of the hydrograph into September and October in years with deep snowpacks. The duration and rate of snowmelt in the study catchments depended not only on the quantity of winter snowfall but also on the elevation, aspect, upstream hydrological network, and topographic complexity of their catchments. Despite having the largest watershed, peak areal discharge in the large, east-facing Ruby Lake watershed was typically lower during snowmelt than at other catchments including Lost Lake, which had considerable terrain with a southern aspect that generated runoff exceeding 40 mm day^{-1} (see figure 19). Factors such as the high altitude of the Ruby catchment, the presence of residual glaciers, or the large capacity of the lake may partially explain these differences. It is also likely that a large percentage of snowmelt recharges a substantial groundwater reservoir and infiltrates into soils and talus fields at Ruby to be slowly released later in the year. Variations in spring and early summer weather exerted considerable effects on the rate of snowmelt. Transient cold fronts with their associated reductions in net shortwave radiative inputs were observed to decrease snowmelt rates in synchrony at all sites throughout the Sierra Nevada by up to 50% even in the peak month of June (see figure 19).

Expressing runoff as the depth of water in millimeters per unit catchment area allows quantitative comparison among the watersheds. In the Emerald Lake watershed, annual average daily runoff ranged from 1.0 mm (1989) to 5.5 mm (1995 and 1998) and averaged 3 mm. Among the other watersheds, average daily runoff from 1990 to 1994 was similar to that for Emerald during that period: 2.3 (Emerald), 2.0 (Pear), 1.7 (Topaz), 1.4 (Ruby), 1.2 (Crystal), 2.3 (Spuller), and 3.6 (Lost). The highest rate of daily areal runoff measured during the study was 94.5 mm d^{-1} at Emerald Lake on October 14, 2009, during an atmospheric river event.

Flow-duration curves (F-D curves) are plots of daily runoff (i.e., outflow discharge per unit catchment area) versus exceedance percentage and illustrate the frequency with which daily runoff values are equaled or exceeded. F-D curves provide information on the hydrologic conditions in watersheds and serve as a means of comparing runoff processes. The Y-intercept represents the peak, daily runoff from a catchment. The slope of the curve is an index of the variability or flashiness of runoff. Slopes less negative are indicative of catchments with low runoff variability, while more negative slopes

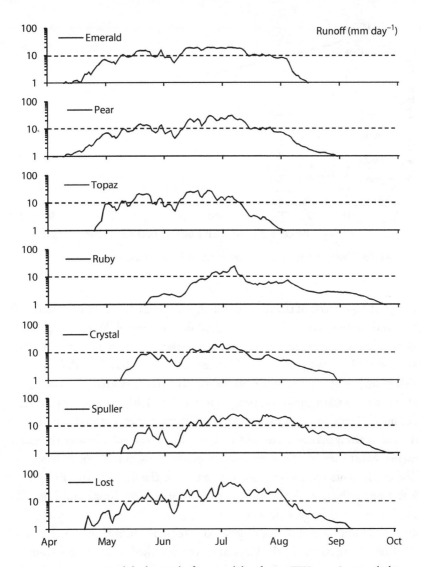

FIGURE 19. Snowmelt hydrographs for seven lakes during WY 1993. Snowmelt discharge is defined as daily discharge greater than 1 mm day^{-1}.

indicate watersheds with higher runoff variability. In general, catchments with larger groundwater storage will have lower peak runoff and flatter F-D curves compared to watersheds with little groundwater capacity.

At Emerald Lake, flow frequency was skewed toward low flows with the 50% flow frequency equal to 0.525 mm d^{-1} (figure 20). Flows above 1 mm d^{-1} were generally confined to the snowmelt period and discharge exceeded

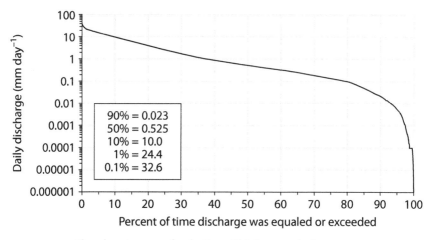

FIGURE 20. Flow-duration curve for the Emerald Lake watershed.

10 mm d^{-1} only 10% of the time. Discharge above 25 mm d^{-1} was exceedingly rare and occurred on only one hundred days during WYs 1984–2016. The slope of the linear regression between exceedance percentage and discharge at Emerald Lake was −0.13 mm d^{-1} exceedance %$^{-1}$. Melack et al. (1998) provide examples of F-D curves for all the watersheds considered. The 1% exceedance for most watersheds was near 20 mm d^{-1}, with higher rates at Lost Lake (34 mm d^{-1}) and lower rates at Ruby and Crystal Lakes (9 mm d^{-1} and 12 mm d^{-1}, respectively). Annual runoff variability was also similar among the catchments. Most sites had flow-duration slopes of > −5 mm d^{-1} exceedance %$^{-1}$. The catchments with slopes nearest zero were the Crystal and Ruby Lake watersheds; those with the most negative slopes were Topaz and Lost. The catchments had large annual variations in runoff and, based on the F-D curves, do not have appreciable groundwater storage, with the exception of the Ruby Lake basin. In the Crystal Lake watershed, water is lost from the catchment via subsurface flow, resulting in low peak runoff and less variable flow since the outflow is normally dry. Groundwater flows at Spuller Lake constitute a small fraction of annual discharge and are not evident from the F-D curve. F-D curve parameters indicate that the Ruby and Lost Lake basins are hydrologically distinct from the others. Snowmelt at Ruby Lake is attenuated and lengthened because the basin is large and at high elevation with sizable groundwater storage and release. In contrast, the Lost Lake catchment is small with south-facing slopes, lies at a lower elevation, has little groundwater storage capacity, and tends to be flashy.

A small percentage of the settled snowpack is lost to evaporation in the central Sierra Nevada (Leydecker and Melack 1999). The typical cumulative evaporative loss from snow varies from 80 mm to 100 mm of water, approximately 7% of the average maximum accumulation. Snowmelt exceeds evaporation because the latent heat of fusion (3.35×10^5 J kg^{-1}) is only 1/80th the latent heat of vaporization (2.5×10^6 J kg^{-1}). In drought years the loss is less but represents a greater percentage of the maximum accumulation, that is, 8% to 13%. During peak snow years the loss decreases to about 4%, although the extended length of the snow season increases the actual amount lost. Most of the loss occurs during the period of snow accumulation when vapor pressure differences are usually favorable for evaporation and conditions of atmospheric instability are found; that is, snow temperatures are warmer than air temperatures. During snowmelt, stable atmospheric conditions and reduced vapor pressure differences, due to the higher vapor content of the warming spring air, reduce evaporation from snow. Hourly measurements of surface temperature with a down-looking radiometer at Mammoth Mountain (Bair, Davis, and Dozier 2018) show that 40% of the time when snow covers the surface, the snow temperature is warmer than the air temperature, indicating that the atmosphere is unstable. In contrast, when the area is snow-free, unstable conditions occur 60% of the time. Total evaporation from the snowpack during snowmelt is typically around 15 mm of water: 2% to 3% of the maximum accumulation during drought years, and less than 1% for above normal snow years. As snowmelt progresses, evaporation from saturated soils and free water becomes appreciable. As water availability diminishes with the gradual drying of the basin, the rate of evaporation decreases.

WATER BUDGETS

Measurements of precipitation and outflow discharge and estimates of evaporation allow calculations of annual water budgets for the Emerald Lake watershed for three decades, for the Tokopah basin for two decades, and for the other six watersheds for four or five years. Snow was the dominant input of water to Emerald Lake and the Tokopah basin in all years except for 2015, when rainfall exceeded spring SWE (about 250 mm) in what is the smallest snowpack on record in the Sierra Nevada. Median annual snowfall was 1,084 mm of SWE at Emerald Lake and 907 mm at the Tokopah basin. Snowfall comprised 33% to 96% of water input, with the highest percentages sometimes measured in the wettest winters, but snowfall would also dominate annual precipitation during moderate to dry years. The largest snowpack at Emerald Lake in our records occurred in 1983 (3,177 mm of SWE) and was larger than the maximum snowpack in our records for the Tokopah basin (2011, 1,851 mm). Rainfall ranged widely at Emerald Lake and the Tokopah basin, with minimum values of 26 mm and 65 mm, respectively, and maximum values of over 400 mm. Rain usually accounted for 10% to 20% of water inputs, but during 2015 (which had the smallest snowpack recorded), rain comprised over half of annual precipitation. Outflow was the major loss of water from the catchments, and median values were 883 mm and 995 mm, respectively. The highest and lowest runoff years were 1983 and 1989 at Emerald Lake (2,122 mm and 366 mm) and 1995 and 2015 at the Tokopah basin (1,954 mm and 493 mm). Outflow exceeded evaporation in all water years. On an annual basis, evapotranspiration comprised between 25% and 35% of annual precipitation.

Residuals are computed as the difference between measured inputs and losses and indicate unmeasured components (e.g., changes in groundwater storage, snow that carries over into the next water year) or measurement errors. If inputs exceed losses the residual is positive; if losses exceed inputs the sign is negative. Errors in the water balances for Emerald Lake and the Tokopah basin were generally small, with median residuals of 14 mm and −12 mm, respectively, indicating our hydrologic measurements were accurate and unbiased. Having an average residual in the water balances near zero demonstrates that the Emerald Lake watershed and the Tokopah basin can be considered hydrologically tight catchments with negligible subsurface loss of water (Kattelmann and Elder 1991; Melack et al. 1998) (figure 21).

The major characteristics of water balances were similar among the other watersheds studied in 1986–87 and 1990–93. As was typical of other catchments, the major input to the Topaz Lake watershed was snow and the major loss, outflow discharge from the snowmelt. Snowfall comprised ~75% to > 90% of basin precipitation. Because of relatively shallow snowpacks and faster recession of snow-covered area, evaporation comprised a larger percentage of water loss at Topaz than at most other catchments. Evaporation accounted for 37% of total precipitation in 1990, 33% in 1991, 43% in 1992, and 19% in 1993. The water balance residual at Topaz was between 10% and 18% of inputs in 1990 and 1991 and less than 5% of inputs in 1992 and 1993.

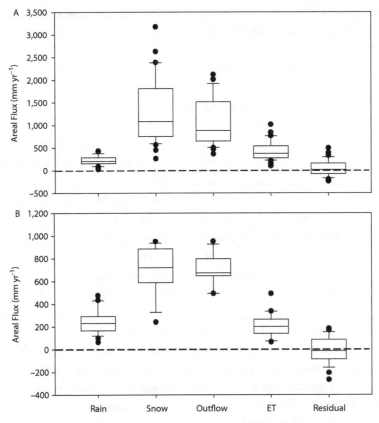

FIGURE 21. (A) Box and whisker diagram of water balances for the Emerald Lake watershed for WYs 1983–2016. (B) Box and whisker diagram of water balances for the Tokopah basin for WYs 1993–2016. ET is estimated evapotranspiration. Positive residuals indicate that inputs were greater than losses; negative residuals occur when losses exceeded inputs. Units: mm yr⁻¹.

The 1993 snowpack completely melted at Topaz, so there was no large, positive residual for this water balance.

In the Tokopah basin, snowfall and outflow accounted for most of the fluxes of water into and out of the basin (figure 21B). Snow constituted 79% of total precipitation, on average, and varied from 47% (2015) to 92% (2004).

In contrast to other watersheds, the Crystal Lake watershed had large residuals and a low percentage of outflow discharge in the water balance that were caused by subsurface flows. Several perennial springs located downslope of the catchment provide evidence of groundwater loss. The shallower slope of the F-D curve also supports this finding. In addition, substantial groundwater flow has been measured in the vicinity of the catchment (Overturf 1991), and springs below Crystal Lake flow throughout summer and autumn when the outlet to Crystal Lake is dry. The Crystal Lake basin and the surrounding area are underlain by volcanic ash and debris conducive to subsurface flows.

In the Ruby Lake basin, snow deposition accounted for 80% to 97% of precipitation, with the highest percentage measured during WY 1993. Evaporation comprised 25% to 40% of water losses. Evaporative losses were greater in 1991 (~46% of inputs) than in 1993 (~16%) owing to persistent snow cover. The residuals for the water budgets at Ruby were small except for an expected positive residual (35% of input) in 1993 caused by unmelted snow that carried over into the next water year. During other years the residual in the annual budget ranged from <1% to 12% of inputs. The small residuals indicate that, despite groundwater inputs to the lake, the watershed as a whole is tight and does not lose much water via subsurface flow.

In the Spuller Lake basin, snow accounted for a high percentage of water input, ranging from ~80% (1992) to greater than 97% (1993) of annual precipitation. Outflow discharge was the largest loss term; during most years, evaporation was typically only half as large as outflow and represented only 13% of water output during 1993. Evaporation decreased in magnitude and as a percentage of total losses when the snowpack was deep and persistent (e.g., 1993). Groundwater inputs to the lake were small on an annual basis but did maintain low outflow between snowmelt runoff periods.

In summary, the hydrology of high-elevation catchments in the Sierra Nevada is dominated by the accumulation and melting of the winter snowpack. The percentage of water lost via outflow was related to the size of the winter snowpack and was greatest following winters with abundant snowfall. In contrast, evaporative losses decreased in percentage when snowpacks were deep owing to the slow recession of snow cover in the basins. The amount of precipitation to the basins that was lost to evaporation ranged from about 10% in wet years to as great as 50% during drought years. The average loss of water to evaporation for the water balances was ~30%, excluding Crystal Lake. Overall, rainfall comprised a small component of the annual water budgets.

FIVE

Watershed Biogeochemistry

Abstract. Atmospheric deposition of chemical species occurs via snow, rain, and dry deposition of particles and gases. Ion concentrations in summer rainstorms are much greater than in winter snow. Outflow solute concentrations are strongly influenced by the accumulation and melting of the seasonal snowpack with base cations and acid neutralizing capacity gradually increasing over winter and diluted by snowmelt runoff, while nitrate pulses during snowmelt are common with lake and watershed processes altering the timing and magnitude of the pulses. The Emerald Lake watershed consumes well over 90% of the substantial loading of hydrogen ion and ammonium and only about half of the nitrate deposition; the catchment generates two to five times more cations than are deposited via atmospheric deposition. While hydrogen deposition declined in the 1980s and 1990s, some lakes still have depressed ANC caused by atmospheric deposition of nitric and sulfuric acid. Interannual variations in solute balances of nitrogen for the Emerald Lake watershed and other Sierra basins indicate that nitrogen retention is high.

Key Words. atmospheric deposition, rain, snow, dry deposition, outflow, solute concentrations, base cations, acid neutralizing capacity, nitrate, snowmelt, Emerald Lake watershed

WATERSHEDS IN THE SIERRA NEVADA INTEGRATE STREAMS and lakes with soils and the atmosphere. Hence we approach watershed biogeochemistry from an ecosystem perspective, as articulated in the classic work by Likens and colleagues (1977). The essence of the approach is an input-output analysis complemented by examination of processes operating within the watershed. Inputs occur via atmospheric deposition modified by weathering reactions and biological activity within the watershed. Outputs occur via stream export plus storage within the watershed. The net difference of inputs and outputs represents the mass balance for the watershed.

We first discuss solute concentrations in atmospheric deposition, how they are measured, and results using three complementary data sets: a time series for the Emerald Lake watershed from water years (WYs) 1984 through 2016, comparative data from several watersheds, and regional data that span the north-south extent of the Sierra Nevada. We then present solute concentrations in the outflow of Emerald Lake and additional data from other watersheds. These results allow the question of whether the inputs or outputs of solutes have changed since the early 1980s.

Second, we examine solute balances for Sierra watersheds utilizing input and output fluxes for major ions and nutrients to determine the extent solute concentrations are altered as they pass through catchments with lakes. These results address the question of whether geochemical and biological processes within the watersheds are sufficient to assimilate inputs of acidic deposition. Because nitrogen and phosphorus are critical nutrients, we develop further discussion of the supply, processing, and export of these two elements.

SOLUTE CONCENTRATIONS IN SNOW AND RAIN

In the Sierra Nevada, atmospheric deposition occurs via snow, rain, and dry deposition of particles and gases. During late autumn, winter, and early spring, the seasonal snowpack accumulates chemical constituents from the snow and dry deposition and releases these materials during snowmelt. During late spring, summer, and early autumn, rain and dry deposition add to the annual deposition. As described in chapters 3 and 4, interannual variations and geographic differences in amounts of snow and rain can be large. Hence we combine our long-term data from the Emerald and Tokopah watersheds with surveys throughout the high-elevation Sierra Nevada.

As concerns about acidic atmospheric deposition in the western United States arose, systematic measurements began in the early 1980s in the Sierra Nevada. The National Atmospheric Deposition Program (NADP) and the California Acid Deposition Monitoring Program (CADMP) established sites at low to moderate elevations on the periphery of the high-elevation Sierra Nevada. Melack and Stoddard (1991) and Blanchard and Tonnessen (1993) summarize initial results from these and other efforts. The Tahoe basin received special attention because atmospheric deposition contributes to eutrophication of the lake (Byron, Axler, and Goldman 1991; Jassby et al. 1994). Intensive measurements in the Emerald watershed began in 1984, and results from the first few years are provided by Williams and Melack (1991a). As these measurements continued and were complemented by sampling from sites throughout the Sierra Nevada and by studies focused on elution of solutes during snowmelt (Harrington and Bales 1998a, 1998b), dry deposition (Bytnerowicz and Olszyk 1988; Bytnerowicz et al.1991; Vicars, Sickman, and Ziemann 2010; Vicars and Sickman 2011), and throughfall (Brown and Lund 1994), a rich literature developed (Melack et al. 1997; Melack, Sickman et al. 1997; Melack et al. 1998; Engle and Melack 1997; Sickman, Leydecker, and Melack 2001; Stohlgren and Parsons 1987; Stohlgren et al. 1991; Fenn et al. 2003). We build on this literature and present the full deposition record for the Emerald Lake watershed in a regional context.

Emissions containing ozone and its precursors (nitrogen oxides, volatile organic compounds, carbon monoxide) from the San Francisco Bay area and the San Joaquin Valley reach the Sierra Nevada (Bytnerowicz et al. 2002; Bytnerowicz et al. 2013; Bytnerowicz et al. 2016; Frączek, Bytnerowicz, and Arbaugh 2003). Forests on the western Sierra slopes are injured by the elevated ozone levels (Peterson et al. 1987; Williams, Brady, and Willison 1977; Carroll, Miller, and Pronos 2003). Concentrations of nitrogenous compounds are generally low during winter when precipitation is derived largely from the Pacific Ocean (Bytnerowicz and Fenn 1996), while dry deposition is an important route for nitrogen loading in the dry season (Fenn, Bytnerowicz, and Liptzin 2012). Western Sierra slopes can experience high levels of ammonia and nitric acid vapor (Bytnerowicz et al. 2016). Pesticides, polybrominated diphenyl ether, polychlorinated biphenyls, and polycyclic aromatic hydrocarbons are also transported into the Sierra Nevada (Bytnerowicz et al. 2016). Desert dust and sea salt aerosols can contribute to formation of cloud ice and precipitation (Rosenfeld et al. 2008), and transpacific transport of aeolian dust from Asia is another source of phosphorus for mid- to high-elevation ecosystems (Vicars and Sickman 2011; Homyak et al. 2014; Aciego et al. 2017). Black carbon comes from California sources, such as diesel engines and wood burning (Chow et al. 2010, 2011), and from as far as Asia (Hadley et al. 2010). Smoke from wildfires is a complex mixture of many compounds (Urbanski, Hao, and Baker 2009) and can be transported long distances to the Sierra (Creamean et al. 2013).

We discuss concentrations of solutes in snow and rain using three complementary data sets: (1) a time series for the Emerald Lake watershed from water years 1984 through 2016; (2) comparative data from the eight watersheds listed in table 1 (Melack et al. 1998) for WYs 1990 through 1994; and (3) regional data from high-elevation sites that span the north-south extent of the Sierra Nevada on both its western and eastern flanks (Melack et al. 1997). Though the data from multiple sites are available for only a subset of years, they put the longer data from the Emerald Lake watershed into a regional context. Because of their role as limiting nutrients, we examine deposition of nitrogen and phosphorus compounds in further detail in the sections on nitrogen and phosphorus biogeochemistry.

Methods

Winter precipitation is defined as that occurring from December through early April, and nonwinter precipitation is defined as that occurring from early April through November. Depth of winter precipitation was measured with intensive snow surveys conducted throughout the watersheds during the time of maximum snowpack (see chaps. 3 and 4). Snow samples for chemical analysis, representing both wet and dry deposition, were collected from snowpits during the snow surveys. Amounts of nonwinter precipitation were measured using tipping bucket rain gauges, and samples for chemical analysis were collected on an event basis from samplers that were triggered to open only during rainfall. In years when rain samplers were not operated at Emerald Lake, we used measurements from nearby NADP stations in Sequoia National Park. Nitrate, chloride, sulfate, acetate, and formate in rain and melted snow were determined by ion chromatography. pH was measured with an electrode designed for use in low conductivity water. Ammonium was assayed colorimetrically, and base cations were determined on an atomic absorption spectrophotometer or by inductively coupled plasma optical emission spectroscopy (ICP-OES). Methodological details are provided by Melack et al. (1997) and Williams and Melack (1991a).

Annual volume-weighted mean (VWM) concentrations were calculated for each solute in winter and nonwinter precipitation. The general equation for calculating annual VWM solute concentrations is given as:

$$C_{vwm} = \frac{\displaystyle\sum_{i=1}^{n} C_i V_i}{\displaystyle\sum_{i=1}^{n} V_i}$$

where:

C_{vwm} = annual volume-weighted mean solute concentration (μequivalent liter^{-1}, μEq L^{-1}),

C_i = solute concentration for event i (μEq L^{-1}),

V_i = water flux associated with event i (liter, L), and

n = the number of snowpit samples or rainstorms during each water year.

Nonwinter dry deposition was computed using the dry-deposition inferential method, which uses airborne chemical concentrations and deposition velocities of SO_2 and HNO_3 to compute deposition rates (Hicks et al. 1991; Meyers et al. 1998). Airborne chemical concentrations were collected by the National Oceanic and Atmospheric Administration (NOAA) Atmospheric Integrated Research Monitoring Network (AIRMoN) at Wolverton Meadow, located 6 km west of the Emerald Lake watershed and at 2,250 m elevation. Methods are fully described by Sickman et al. (2001). After the closure of the Wolverton Meadow station we used dry deposition measurements from the EPA's Clean Air Status and Trends Network (CASTNET) stations in Sequoia and Yosemite National Parks to estimate dry deposition at Emerald Lake and the Tokopah basin.

Emerald Lake Watershed

Snow chemistry in the Sierra Nevada is very dilute, with most concentrations only a few μequivalents per liter (figure 22). The presence of acidic anions, nitrate and sulfate, cause the pH of snow to be slightly below the expected level of 5.65 if only carbonic acid contributed to acidity; mean pH of snow at Emerald Lake is 5.53. The dominant cation in snow is typically either hydrogen ion or ammonium, with the dominant anion being nitrate; ammonium acts as a base in the snow, neutralizing some of the acidity. On a molar basis, sodium is the dominant base cation followed by calcium and then magnesium or potassium. Between 1985 and 2016 several changes in the chemistry of snow, rain, and annual precipitation at Emerald Lake occurred based on 4-year moving averages of annual volume-weighted means (figures 22–24).

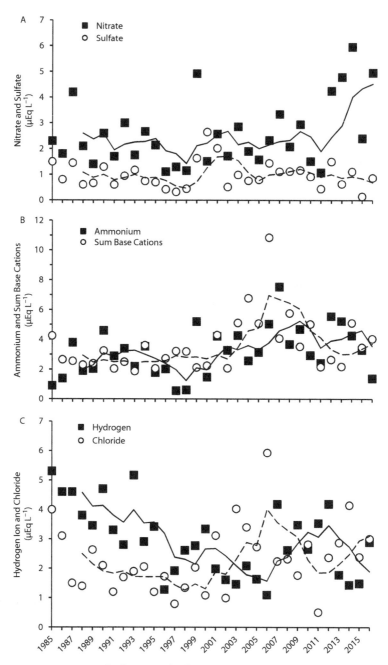

FIGURE 22. Annual volume-weighted mean nitrate and sulfate (A), ammonium and sum of base cations (B), and hydrogen ion and chloride (C) for the spring snowpack measured at Emerald Lake in the period 1985–2016. Solid and dashed lines are 4-year moving average annual volume-weighted means.

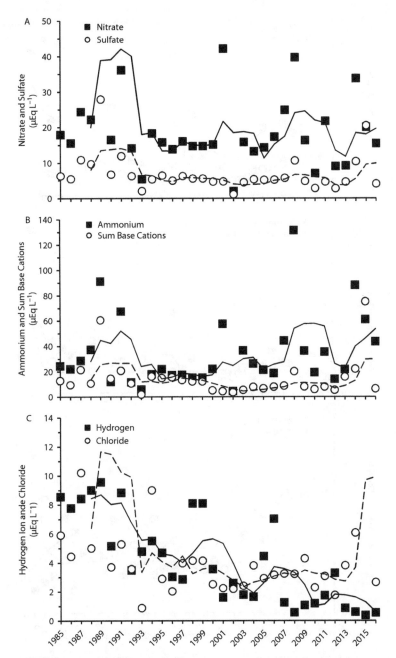

FIGURE 23. Annual volume-weighted mean nitrate and sulfate (A), ammonium and sum of base cations (B), and hydrogen ion and chloride (C) for nonwinter precipitation (rain and snow falling from ~May to November) in 1985–2016. From 1985 to 2001, all measurements were made at Emerald Lake. After 2001, we used a combination of Emerald Lake measurements and data from the Giant Forest NADP station to compute the volume-weighted mean values. Lines as in figure 22.

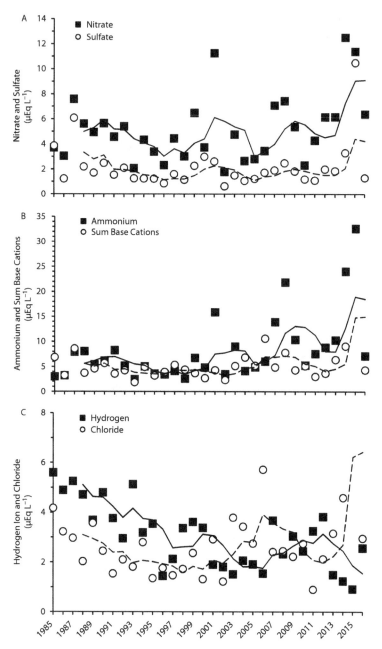

FIGURE 24. Annual volume-weighted (snow plus nonwinter precipitation) mean nitrate and sulfate (A), ammonium and sum of base cations (B), and hydrogen ion and chloride (C) during WYs 1985–2016.

Concentrations in winter snow tended to rise and fall in response to changes in the amount of winter snowfall. Generally lower concentrations were detected in years with high snowfall, perhaps due to dilution of winter dry deposition by large snowstorms. Notable features or trends in the 32-year record include elevated nitrate levels in snow during the 2012–16 drought and higher base cation and ammonium levels in snow in the 2000s relative to the 1980s and 1990s. Hydrogen ion in snow gradually declined from the 1980s through 2007 and then rose slightly. Our data suggest that air quality improvements resulting from the 1970 Clean Air Act and Amendments (1977, 1990) have reduced the rate of acid loading to the Sierra Nevada (Heard et al. 2014). Recent increases in nitrogen deposition are troubling and indicate that more should be done to improve air quality, especially since current rates of N deposition exceed the critical load in many areas of California, including the Sierra Nevada (Heard and Sickman 2016).

Nonwinter precipitation falls as rain and snow from May through November. Precipitation during spring and late autumn usually derives from the midlatitude Pacific Ocean and tends to be only slightly enriched in ions relative to winter snow. In contrast, summer and early autumn rain is often derived from storms formed from local air masses that are influenced by fossil fuel burning and agricultural activities. Ion concentrations in small summer rainstorms can be as much as one hundred times greater than in winter snow (see figure 23). The dominant ions are ammonium and nitrate. The major sources of ammonium in the Sierra Nevada are fertilizer applications and confined animal feeding operations, which produce large amounts of NH_3 that is advected into the high-elevation Sierra. Dry deposition of ammonium and nitrate is also elevated in the Sierra Nevada, and nonwinter wet and dry N deposition is typically of the same magnitude.

The ionic composition of annual nonwinter precipitation varies more than that of snow because of the greater chemical variability among the rain events. For example, in years with abundant spring and late autumn storms, nonwinter precipitation is more dilute. In years with little spring or autumn precipitation but ample rains in the summer monsoon season, ion concentrations of nonwinter precipitation are relatively high. Because of this greater interannual variability, it is difficult to detect changes in nonwinter precipitation chemistry that are related to long-term trends in air quality. Through the 1980s and mid-1990s we noted a decline in sulfate, followed by little change until the 2012–16 drought when sulfate concentrations increased (see figure 23). Nitrate concentration prior to the year 2000 typically ranged from

15 to 20 μEq L⁻¹, but later in the record, interannual variability increased, with peak values exceeding 30 μEq L⁻¹. Ammonium levels in the late 2010s were also elevated relative to earlier years. Hydrogen ions in rain at Emerald Lake gradually declined until 2007 and then stabilized.

Despite the relatively high concentrations of the ions in nonwinter precipitation, the VWM concentrations of major ions in annual precipitation (sum of winter snow and nonwinter precipitation) is similar to winter snow deposition (see figure 24). Between 1985 and 2016, winter snowfall in the Emerald Lake watershed averaged 1,232 mm, while nonwinter precipitation was 213 mm (~15% of annual precipitation); thus the chemistry of snow is the dominant factor controlling annual VWM chemistry of most ions. Temporal trends in snow and rain chemistry were detectable in the annual precipitation chemistry. Most notable was the increase in nitrate and ammonium in the late 2010s and the decline in hydrogen in the first twenty-five years of the record.

While snowfall dominates atmospheric input of water to the Sierra Nevada, other forms of atmospheric deposition were larger sources of loading for acids and nutrients at Emerald Lake (figure 25). Dry deposition during nonwinter periods was the dominant input of hydrogen ion and nitrate at Emerald Lake, contributing on average more than 50% of the loading. In contrast, an average of 60% of the ammonium deposition was contributed by nonwinter rain, and rain was the dominant route for sulfate deposition. Winter snowfall contributed the majority of the annual loading of chloride (73%) and the base cations (51%–66%). These observations are relevant to predictions of future climate in the Sierra Nevada that suggest reduction in winter snowfall (see chapter 7). If winter snowfall declines, then atmospheric base cation loading will be reduced.

Eight Watersheds

Comparison of precipitation chemistry at Emerald Lake with a larger set of sites in the Sierra Nevada shows that the Emerald Lake watershed is typical of other high-elevation catchments and that long-term trends noted at Emerald Lake are likely occurring throughout the Sierra Nevada.

Among the eight watersheds, potassium and organic anions (acetate and formate) were the most variable solutes in snow, probably because of concentrations near the detection limits (table 2). The least variable solute was hydrogen ion, and mean pH of snow was 5.42 (range, 5.57–5.25). Average

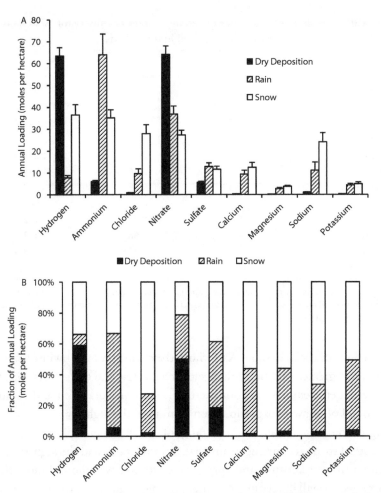

FIGURE 25. Snow, rain, and dry deposition to the Emerald Lake watershed expressed as moles per hectare for each solute (A) and as fraction of annual loading (B).

nitrate concentrations in snow (2.4 μEq L^{-1}) slightly exceeded sulfate (2.0 μEq L^{-1}). After hydrogen ion and ammonium, the next most abundant cations were calcium and sodium; the average calcium concentration in Sierra snow was about 30% greater than the average sodium level.

In contrast to snow, nonwinter precipitation measured in the eight watersheds was relatively rich in solutes (table 3). Ammonium, nitrate, sulfate, and hydrogen ions had greater concentrations than in winter snow. Total cations and anions were, respectively, 54.8 and 56.4 μEq L^{-1}, about five times greater

TABLE 2 Average of volume-weighted mean snow chemistry for WYs 1990
through 1994 in eight watersheds

Solute	Mean	C.V.	Range
Hydrogen	3.8	0.21	2.7–5.6
Ammonium	2.7	0.38	0.9–5.5
Chloride	1.7	0.51	0.6–4.4
Nitrate	2.4	0.33	1.2–4.5
Sulfate	2.0	0.26	1.2–3.0
Calcium	1.7	0.46	0.6–4.2
Magnesium	0.5	0.43	0.3–1.4
Sodium	1.3	0.48	0.5–4.0
Potassium	0.5	0.90	0.1–2.3
Acetate	0.6	0.68	0.0–2.2
Formate	0.5	0.66	0.0–1.4
Deposition	1,027	0.60	501–3,017

NOTE: Arithmetic means, coefficient of variation (C.V.), and the range of values measured for the 36 catchment-years of record for each constituent (μEq L^{-1}); deposition is precipitation in millimeters.
SOURCE: From Melack et al. 1998.

than concentrations in snow. As a whole, the chemistry of nonwinter precipitation had greater variability than snow chemistry, and the solutes with the greatest coefficients of variation were ammonium, potassium and sodium. Mean pH of nonwinter precipitation from 1990 through 1994 was 4.93 and ranged from 4.59 to 5.66. The large range of nonwinter precipitation is due to the mix of storms occurring in any year. Summer rains had the highest solute concentrations; autumn and spring precipitation was more dilute. These storms were small in relation to winter precipitation, and in total, mean nonwinter precipitation represented only ~10% (117 mm) of mean annual winter deposition.

Despite considerable interannual variability in precipitation chemistry and quantity, annual wet deposition was fairly consistent from year to year and among the eight catchments. Annual wet deposition is more similar to winter snowfall than nonwinter precipitation because winter snowfall exceeds nonwinter precipitation in the Sierra Nevada. Solutes consist primarily of hydrogen ion, ammonium, nitrate, and sulfate with some chloride and calcium. Together these solutes comprise 63% of all ions in solution. Other constituents, potassium, sodium, magnesium, and organic anions, were found at lower concentrations. The mean annual pH for wet deposition was

TABLE 3 Average of volume-weighted mean chemistry of nonwinter precipitation
for WYs 1990 through 1994 in eight watersheds

Solute	Mean	C.V.	Range
Hydrogen	11.7	0.61	2.2–25.8
Ammonium	23.4	0.70	5.8–67.4
Chloride	4.2	0.53	0.9–10.0
Nitrate	20.7	0.46	5.4–45.8
Sulfate	15.1	0.45	4.1–39.8
Calcium	10.4	0.59	1.0–22.9
Magnesium	2.3	0.62	0.2–5.6
Sodium	4.6	0.66	0.8–15.5
Potassium	2.4	0.68	0.5–6.1
Acetate	7.3	0.63	2.0–20.6
Formate	9.1	0.58	2.6–19.6
Deposition	117	0.51	8–236

NOTE: Arithmetic means, coefficient of variation (C.V.), and the range of values measured for the 36 catchment-years of record for each constituent (μEq L^{-1}); deposition is precipitation in millimeters
SOURCE: From Melack et al. 1998.

5.3 and ranged from 5.1 to 5.5. Ammonium and nitrate were the most abundant cation and anion in solution, with mean concentrations of 5.1 μM and 4.6 μEq L^{-1}, respectively. The average sulfate concentration in wet deposition was slightly lower, 3.6 μEq L^{-1}, and ranged from 1.8 to 9.9 μEq L^{-1}. The mean concentration of acetate and formate in wet deposition was 1.4 μEq L^{-1}. Mean annual precipitation for the study was slightly over one meter (1,144 mm) of water equivalence and varied by nearly a factor of five.

Regional Surveys and Interseasonal Variability

In previous sections we emphasized annual volume-weighted means for precipitation chemistry. Here we investigate the variability of individual rain events and snow samples across a set of monitoring stations spanning both dry and wet years. The quantity and chemical composition of precipitation throughout the Sierra Nevada were monitored at multiple stations during the period 1990 through 1995, the only period with such detailed data (readers are encouraged to read technical report A932–081 to the California Air Resources Board by Melack, Sickman et al. 1997, available at https://ww3.arb.ca.gov/research/). Winter precipitation ranged from about 250 mm to

3,000 mm. Nonwinter precipitation events ranged in size from trace amounts to over 100 mm, and annual amounts of nonwinter precipitation varied from 10 mm to 200 mm.

Solute concentrations in snow were similar among the six years of study and among the sampling stations. Individual samples from the spring snowpack had pH typically between 5.3 and 5.6 (hydrogen ion concentration, 2 to 4 μEq L^{-1}), and ammonium and nitrate concentrations were usually 1.5 μEq L^{-1} to 4.5 μEq L^{-1}. Sulfate concentrations ranged from 1.0 μEq L^{-1} to 3.0 μEq L^{-1}. Organic anions (acetate and formate) were usually found at low levels (< 0.5 μEq L^{-1}).

Mean pH of summer rain during the study period was 4.65; the lowest pH measured was 3.86. The range of pH in spring and autumn storms was 4.4 to 6.0, with mean values of 5.05 for spring storms and 5.00 for autumn storms. Spring storms had a mean ammonium concentration of 18.9 μEq L^{-1} (n = 67) and a range of 1.1 μEq L^{-1} to 80 μEq L^{-1}. Mean ammonium concentration in summer storms was 36.5 μEq L^{-1} (n = 365) and ranged up to about 160 μEq L^{-1}. Mean ammonium concentration in autumn storms was 28.1 μEq L^{-1} (n = 82) and ranged from 1.0 μEq L^{-1} to 160 μEq L^{-1}. Mean nitrate concentration in summer rain was 36.4 μEq L^{-1}, and levels over 40 μEq L^{-1} were common. Nitrate concentrations in autumn precipitation averaged 22.3 μEq L^{-1}, and spring storms had the lowest nitrate levels, averaging 17.3 μEq L^{-1}. Summer rains had sulfate levels that averaged 26.8 μEq L^{-1}, and sulfate was significantly lower in spring and autumn precipitation, with means of 16.4 μEq L^{-1} and 14.0 μEq L^{-1}, respectively. Mean concentrations of acetate and formate in summer rains were about 12 μEq L^{-1} to 14 μEq L^{-1}, and mean values measured in spring and autumn precipitation were on the order of 3 μEq L^{-1} to 5 μEq L^{-1}, respectively. High levels of organic anions occurred in small (less than 5 mm) summer storms and exceeded 60 μEq L^{-1}.

Mean annual hydrogen ion deposition among the sites ranged from 24.1 Eq ha^{-1} to 60.9 Eq ha^{-1}. At the majority of stations and during most years, deposition of hydrogen ion was greater in winter than in nonwinter seasons, but these estimates do not include dry deposition. Average annual ammonium loading varied from 21.6 Eq ha^{-1} to 58.4 Eq ha^{-1}. Mean nitrate deposition ranged from 22.4 Eq ha^{-1} to 53.7 Eq ha^{-1}, and average annual sulfate deposition ranged from 16.9 Eq ha^{-1} to 44.5 Eq ha^{-1}. Average annual deposition of calcium varied from 14.6 Eq ha^{-1} to 38.6 Eq ha^{-1}, and sodium deposition ranged from 4.8 Eq ha^{-1} to 27.8 Eq ha^{-1}. Deposition of magnesium and potassium on an annual basis ranged from 2.9 Eq ha^{-1} to 11.0 Eq ha^{-1}. The amount of acetate and formate deposition at most sites was similar and ranged from 2.2 Eq ha^{-1} to 19.1 Eq ha^{-1}, respectively.

We used these data on solute concentrations to investigate geographic differences by dividing the Sierra Nevada into subunits based on latitude, altitude, and the location of our sampling stations (Melack, Sickman et al. 1997). Snow water equivalence data from the California Cooperative Snow Survey (CCSS) obtained near April 1 each year were combined with snow chemistry to calculate solute deposition in each region. Snow chemistry from Pear Lake, Topaz Lake, Ruby Lake, and Crystal Lake was also used in these calculations (Melack et al. 1998). Wet solute loading in most cases was greatest above 2,500 m. Latitudinal trends in solute loading loosely followed the same patterns as precipitation quantity. During 1991, 1992, and 1993 hydrogen ion deposition decreased from north to south. Ammonium deposition was highest in the 39° to 38°30' region during 1993 and 1991, the two wettest winters. During the relatively dry years of 1990 and 1992, deposition was greatest in the northernmost subunits, although we lack dry deposition measurements. Regional loading of nitrate had patterns fairly similar to ammonium, as did chloride. Sulfate and base cation loading had peaks in loading in the central Sierra, with lesser amounts north and south in 1991 and 1992. During 1993 sulfate and base cation deposition was greatest in the 39° to 38°30' subunit and decreased to the south. Sulfate deposition in the two driest years, 1990 and 1992, and base cations loading in 1990 varied little among the regions but tended to be highest in the northern and southern regions.

Differences in solute loading are explained by two factors: variability in precipitation quantity and variability in precipitation chemistry. These regional deposition data indicate that the southern Sierra Nevada receives more loading of ions per unit of precipitation compared to the northern Sierra. During winter snowfall, the relative importance of these two factors is evident when comparing solute loading to snow water equivalence. For certain ions there is a strong linear relationship between winter loading and SWE. For instance, 83% of the variability in sulfate loading in the Sierra Nevada is explained by SWE, and differences in SWE account for 70% to 75% of the variability in winter loading of sodium and chloride. In contrast, less than 50% of the variability in organic anions and potassium was explained by SWE. For nitrate, ammonium, and the other base cations, SWE explained approximately 50% of the variability in loading. These data indicate a common source, likely the Pacific Ocean, for most of the sulfate, sodium, and chloride deposited in the Sierra Nevada during the winter. Data for the other ions suggest that local sources of solutes exist and vary from subregion to subregion.

RELEASE OF SOLUTES FROM SNOW

As snow melts, solutes are released and eventually reach streams and lakes. While snowmelt depends on the energy balance of the snowpack and the process can be modeled (see chapter 3; see also Marks and Dozier 1992; Cline, Bales, and Dozier 1998; Marks et al. 1999), the release of solutes is more complex (Harrington and Bales 1998b). Melt-freeze cycles can cause the outer coating of snow crystals to become enriched with solutes. Rosenthal, Saleta, and Dozier (2007) examined midwinter Sierra snow using an environmental scanning electron microscope and observed ringlike filaments at grain boundaries. Energy dispersive X-ray spectrometry detected elements typical of soluble impurities in Sierra snow. These coatings can be "washed" from the snow during the first releases of water from the snowpack and can result in elevated solute levels (Colbeck 1981; Williams and Melack 1989, 1991a; Harrington and Bales 1998a). This preferential elution can lead to substantially greater concentrations of solutes, such as nitrate, than in the snowpack and cause episodic acidification (Stoddard 1995). We discuss nitrate pulses further in subsequent sections.

OUTFLOW SOLUTE CONCENTRATIONS

Seasonal and Interannual Patterns

In this section we characterize the seasonal and interannual patterns in major solute concentrations in streams at the outlets of lakes and discuss similarities and differences among the catchments. The data required to detect a significant temporal trend or pattern depend on the frequency of sampling, natural variability of the parameter of interest, and the magnitude of the change. Our best data sets for detecting changes are the outflow time series from Emerald Lake. Hence we first present our time series for the Emerald Lake outflow, followed by consideration of seasonal patterns and interannual differences among other watersheds.

In parallel with measurements of atmospheric deposition in the early 1980s, sampling of the streams exiting high-elevation watersheds in the Sierra Nevada was initiated. Collections of stream water were made as frequently as daily to weekly during snowmelt and into summer and then biweekly to bimonthly from November to April. The major ions in streams in the Sierra Nevada are the cations, calcium, magnesium, sodium, potassium, and hydrogen (recorded as pH); and the anions, nitrate, sulfate, chloride, and bicarbonate (recorded as acid neutralizing capacity [ANC]). Further information on sampling, sample processing, and chemical analyses is provided by Williams and Melack (1991b) and Melack et al. (1998).

Emerald Lake Watershed

Outflow solute concentrations at Emerald Lake are strongly influenced by the accumulation and melting of the seasonal snowpack (figure 26). Two general patterns are discernible: ions that gradually build up over the winter and are diluted by snowmelt runoff (base cations and ANC) and ions that have a pulse-dilution pattern (nitrate, sulfate, and hydrogen). Base cations and ANC are correlated, and their concentrations increase over the winter, reaching a peak in March. As snowmelt runoff initiates in April, ANC and base cation concentrations decline, reaching minima during the peak of runoff in June. Outflow ANC and base cations declined on average by 30% to 35% in response to snowmelt. In contrast, sulfate, hydrogen ions, and especially nitrate increase in concentration on the rising limb of the snowmelt hydrograph. Small pulses of sulfate and hydrogen ion occur in April, but the larger pulse of nitrate does not peak until May or June. Based on stable isotope analyses, the sulfate (J. Sickman, unpublished data) and nitrate (Sickman, Leydecker et al. 2003) pulses originate predominantly from flushing of S and N from catchment soils and to a lesser extent from preferential elution from the snowpack. These sources are increasingly exhausted as runoff peaks in June and July and concentrations decline rapidly. Of the major ions, sulfate concentrations vary in a narrow range despite discharge changing by several orders of magnitude. The processes that regulate sulfur biogeochemistry in the Sierra Nevada are poorly understood. Outside of regions with volcanic activity (e.g., Crystal Lake), most bedrock in the Sierra Nevada has little sulfur, and the regulation of sulfate concentrations in surface water is likely controlled by hydrologically mediated adsorption and desorption of sulfate from exchange sites on soils (Clow, Mast, and Campbell 1996).

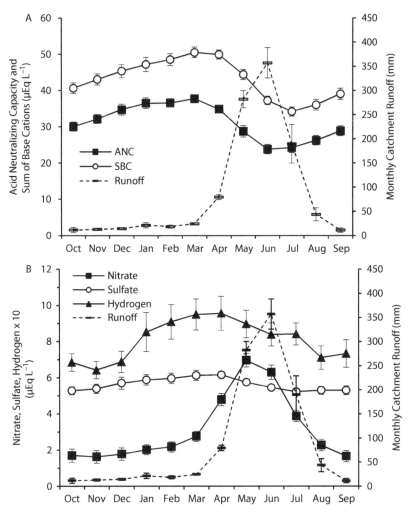

FIGURE 26. Average monthly outflow chemistry vs. average monthly discharge for Emerald Lake during 1985–2016. Vertical bars denote standard errors.

The winter buildup and subsequent snowmelt dilution of ANC and base cations produces concentration-discharge relationships having a clockwise hysteresis: (1) decreasing concentrations with increased flow on the rising hydrograph limb; (2) followed by low, nearly constant, concentrations on the falling limb; and (3) gradual recovery to pre-snowmelt concentrations beginning in autumn (figure 27). ANC depression was greatest during years with deep snowpack and large snowmelt runoff, such as 1995.

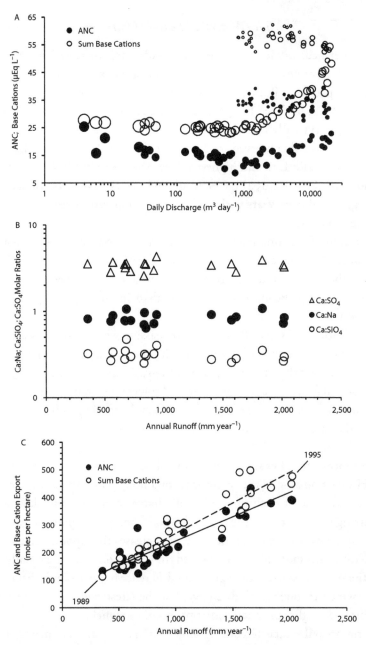

FIGURE 27. Temporal dynamics of ANC, sum of base cations (SBC), and silica in the Emerald Lake watershed. (A) Discharge vs. concentration relationships for the snowmelt season in 1995 (increasing size of the markers represents forward movement through time). (B) Volume-weighted mean molar ratios of cations and silica plotted against annual runoff; in these plots there was a significant relationship between the molar ratio and runoff. (C) Linear regressions between annual watershed runoff and outflow flux of ANC (r^2, 0.83) and SBC (r^2, 0.90).

In all watersheds studied, ANC declined as discharge increased during snowmelt, with minimum ANC occurring at or near peak runoff. A good example of this pattern is illustrated by Emerald Lake's outflow in 1993 and 1994. Outflow ANC declined by 25% to 80% as a result of snowmelt, with minimum values typically in the range of 15 μEq L^{-1} to 30 μEq L^{-1}. The lowest values were usually observed in the Lost Lake and Pear Lake outflows (10 μEq L^{-1} to 15 μEq L^{-1}), and highest values were measured in Crystal Lake's outflow (50 μEq L^{-1} to 65 μEq L^{-1}). In all catchments ANC depression tended to be greater during years with high runoff compared to years with below average snowfall.

The sum of calcium, magnesium, sodium, and potassium and dissolved silica in outflows decreased in concentration during snowmelt, with minima at or near the time of peak runoff. Base cations and silicate were significantly lower ($p < 0.05$) during peak discharge than in the beginning phases of snowmelt.

Fluctuations in streamflow caused by passing weather systems are common during the months of April through June, and a good example of this occurred during the snowmelt period of 1993. In late May and early June, a series of cold fronts brought light snow and colder temperatures to California. Snowmelt runoff declined during these intervals, causing an increase or leveling off of base cations and silicate concentrations in the outflows of nearly all catchments.

In contrast to other solutes, it is difficult to describe a consistent pattern of pH variation among the catchments or among years. The most common pattern was one of declining pH as discharge increased, with lowest pH occurring sometime near the peak of runoff, though this pattern was often interrupted by transient pH increases or decreases that appear unrelated to discharge. Minimum pH values ranged from 5.5 to 6.1 and were usually near 5.8. There was no statistically significant difference in pH from the beginning of snowmelt through peak discharge, but the latest stage of snowmelt had significantly higher pH than earlier phases of snowmelt.

A nitrate pulse occurred during snowmelt in nearly all catchments; Topaz Lake was the exception. Streams from Emerald Lake and the Tokopah basin had nitrate maxima between 6 μEq L^{-1} and 8 μEq L^{-1} during most years. At Pear and Ruby Lakes peak nitrate levels ranged from 7 μEq L^{-1} to 12 μEq L^{-1}.

Highest snowmelt nitrate levels were observed in the outflow from Spuller Lake, 13 ± 3 μEq L^{-1}. The pulse was muted or absent in waters outflowing from Lost and Crystal Lakes, where peak values ranged from about 0.5 μEq L^{-1} to 2 μEq L^{-1}. Inflowing waters to Lost and Crystal Lakes did, however, indicate a nitrate pulse; peak concentrations ranged from 5 μEq L^{-1} to 10 μEq L^{-1}. Thus in-lake biological processes consumed much of the nitrate delivered from the Lost and Crystal Lake watersheds.

At Topaz Lake during WYs 1988, 1991, and 1993, unusually high nitrate concentrations were measured in autumn and winter and in the spring prior to snowmelt; levels were 40 μEq L^{-1} to 175 μEq L^{-1}, much higher than peak nitrate values of less than 20 μEq L^{-1} observed at other catchments. Sickman, Leydecker et al. (2003) hypothesized that these high nitrate concentrations resulted from senescence of plants in the extensive meadows surrounding the lake, which increased microbial nitrogen mineralization and nitrification. Sickman, Melack, and Stoddard (2002) reported springtime nitrate peaks up to 38 μM for other talus-dominated catchments at high elevation. Thus a nitrate pulse appears to be a nearly universal biogeochemical feature in Sierra watersheds, and lake and watershed processes appear to alter the timing and magnitude of the pulse.

Seasonal patterns in sulfate concentration were often qualitatively similar to nitrate patterns, but the magnitudes of the changes were smaller. In some cases, dilution throughout snowmelt was observed; for example, Spuller Lake outflow sulfate declined from 15 μEq L^{-1} to 25 μEq L^{-1} before snowmelt to about 5 μEq L^{-1} at peak runoff. The lower portions of the Spuller watershed are composed of ancient tuffaceous lakebeds of volcanogenic origin and likely weather more sulfate than the more typical granitic bedrock. The magnitude of sulfate decline varied considerably among the watersheds, and reductions of less than 1 μEq L^{-1} to 2 μEq L^{-1} were observed in many cases. In these situations, sulfate declined by less than 30% from concentrations prior to snowmelt.

Multi-Watershed Interannual Comparison

Annual volume-weighted mean concentrations were calculated for each solute in outflow discharge in a manner analogous to that for precipitation, as described in Melack et al. (1998). The average pH for outflows during the study was 6.05 and ranged from 5.6 to 6.7 (table 4). The pH of outflow

Solute	Mean	C.V.	Range
Hydrogen	0.9	0.56	0.2–2.5
ANC	36.2	0.42	17.4–73.1
Ammonium	0.2	0.91	0.0–0.8
Chloride	2.9	0.26	1.1–4.4
Nitrate	3.2	0.64	0.2–7.8
Sulfate	7.4	0.30	4.6–12.1
Calcium	28.2	0.39	14.5–50.5
Magnesium	4.7	0.59	2.8–12.5
Sodium	11.9	0.31	6.2–22.3
Potassium	3.7	0.37	2.0–7.6
Silicate	34.7	0.58	15.4–72.9
Outflow	787	0.62	389–2,359

NOTE: Arithmetic means for each major constituent ($\mu Eq\ L^{-1}$), coefficient of variation (C.V.), and the range of values measured for 36 catchment-years of record. Runoff is the equivalent water depth of outflow discharge in millimeters. Data from Crystal Lake were not included in the runoff statistic because of large subsurface water loss. Acetate and formate are undetectable in the surface waters of the Sierra Nevada. Silicone concentration in μM.

SOURCE: From Melack et al. 1998.

from the Lost Lake catchments was significantly ($p < 0.05$) lower than the pH at Crystal, Ruby, and Spuller Lakes. Catchments with the lowest ANC (15 to 30 $\mu Eq\ L^{-1}$) were Lost, Pear, and Emerald Lakes. Watersheds with ANC in the range of 30 to 50 $\mu Eq\ L^{-1}$ include Topaz and Spuller Lakes and the upper Marble Fork. Crystal and Ruby Lakes had ANCs above 50 $\mu Eq\ L^{-1}$. Sulfate was the most consistent solute among the catchments and across water years with average VWM sulfate concentration of 7.4 $\mu Eq\ L^{-1}$. Nitrogen transformations within the catchments had a large effect on the levels of ammonium and nitrate in outflowing waters, as discussed further below. The average ammonium concentration was 0.2 $\mu Eq\ L^{-1}$ and ranged from 0.0 to 0.8 $\mu Eq\ L^{-1}$. Crystal, Lost and to some extent Topaz (excepting autumn and winter pulses), also had low levels of nitrate in outflow streams (about < 1 $\mu Eq\ L^{-1}$). Other catchments had nitrate concentrations of 3 to 8 $\mu Eq\ L^{-1}$.

Calcium and sodium were the major cations in outflows (table 4). Calcium ranged from 20 $\mu Eq\ L^{-1}$ to 50 $\mu Eq\ L^{-1}$, with the highest levels measured in catchments in the eastern Sierra Nevada. The average sodium concentration was less than half that of calcium: 11.9 μEq L^{-1}. An exception was Crystal Lake, which was enriched with sodium and magnesium relative to calcium, reflecting the volcanic origin of its surrounding soils. The concentrations in this catchment ranged from 19.4 μEq L^{-1} to 22.3 μEq L^{-1} for sodium and were near 12 μEq L^{-1} for magnesium. Concentrations for magnesium and potassium were relatively low, with means of 3.7 μEq L^{-1} and 4.7 μEq L^{-1}, respectively. Silicate concentrations ranged from 15.4 μM to 72.9 μM, with a mean of 34.7 μM.

SOLUTE BALANCES

Precipitation intercepted by watersheds undergoes chemical alterations before exiting in streams. Chemical weathering of granitic minerals, cation exchange in soils, and biotic activity are some of the processes responsible for these alterations. Analyses of watershed solute balances utilize comparisons between input and output fluxes for major ions and nutrients to determine the extent that solute concentrations are altered as they pass through a catchment. Since the Sierra Nevada is diverse with respect to geology, soils, precipitation, vegetation, and topography, we present solute balances for seven headwater catchments and for the upper Marble Fork of the Kaweah River (Tokopah basin). We compare and contrast the solute balances and examine the chemical alterations that occur in Sierra catchments with an emphasis on the Emerald Lake watershed. Using these data we examine the influence of various factors controlling surface water chemistry in Sierra watersheds on both seasonal and annual time scales.

Solute loading in the Sierra Nevada is the sum of solutes contributed by wet and dry atmospheric deposition. Since estimates of nonwinter dry deposition were only available for the Emerald, Pear, and Topaz watersheds and depended on measurements in a forested area at about 2,000 m elevation, solute loading for the comparative results presented here do not include nonwinter dry deposition. Loss is the transport (export) of dissolved and particulate material in the outflow streams of the catchments; we consider only dissolved constituents, with the exception of nitrogen in a few years. Hydrological information relevant to calculation of solute balances is presented in chapter 4.

Nonwinter and winter solute loading were calculated by multiplying VWM solute concentrations by the amount of precipitation and then normalizing the product to watershed area:

$$L = \frac{C_{VWM} \times P}{A}$$

where:

L = nonwinter or winter solute loading (equivalents or moles ha^{-1}),

C_{vwm} = VWM solute concentration in nonwinter or winter precipitation (equivalents or moles m^{-3}),

P = volume of nonwinter or winter precipitation deposited in catchment (m^3), and

A = area of catchment (ha).

Annual solute export E_A was calculated as the product of outflow discharge and VWM solute concentration normalized to watershed area:

$$E_A = \frac{C_{VWM} \times Q_A}{A}$$

where:

E_A = annual solute export (equivalents or moles ha^{-1}),

C_{vwm} = annual VWM solute concentration in outflow (equivalents or moles m^{-3}),

Q_A = volume of annual outflow discharge (m^3), and

A = area of catchment (ha).

Watershed solute yield (Y_A) was computed as:

$$Y_A = E_A - L_A$$

where:

E_A = the annual flux of dissolved constituents in outflow discharge, and

L_A = the annual solute loading

Yields are expressed in moles per hectare of catchment area (mole ha^{-1} yr^{-1}). Solute yields greater than zero mean that losses exceeded inputs, and the catchment is a source for that particular solute. Conversely, yields less than zero indicate that inputs exceeded losses and that the watershed is a net sink. Yields near zero indicate the solute behaves conservatively with respect to watershed processes. Fluxes were normalized to catchment area so that comparisons could be made among sites.

Sources of Error in Solute Balances

Four types of errors can occur in solute balance estimates: (1) volumetric hydrologic errors, (2) chemical analytical errors, (3) sampling errors, and (4) systematic errors or bias. Volumetric errors are the uncertainty in estimates of winter and nonwinter precipitation volume and outflow discharge. Since solute fluxes are the product of water volumes and VWM concentrations, errors in measuring precipitation or streamflow will affect the uncertainty of the fluxes. Another source of error is introduced during chemical analyses of precipitation and stream samples. Missed precipitation and low-frequency stream sampling will introduce errors into VWM chemistry in addition to analytical error. Several systematic errors were also identified, the most important of which were (1) overestimates of outflow discharge at catchments without weirs, (2) underestimates of precipitation at catchments without nearby weather stations (e.g., Lost Lake), and (3) lack of nonwinter dry deposition measurements. Detailed estimates of all the solute balance errors are provided by Melack, Sickman et al. (1997) and Melack et al. (1998).

Errors in estimating precipitation depended on the timing of storms and whether precipitation fell as snow or rain. We estimated that the measurements of winter snow accumulation had an uncertainty of ± 5%. Errors in nonwinter precipitation volume depended on the timing and form of the precipitation, and, because of logistical factors, some catchments had greater error than others. In the Crystal, Emerald, Marble Fork, Pear, and Topaz watersheds, nonwinter precipitation volume had an error of ± 9%; uncertainty in nonwinter precipitation at the Lost, Ruby, and Spuller watersheds was ± 14%. Errors in discharge measurements varied from site to site and from year to year. On an annual basis, uncertainty in outflow ranged from ± 5% to ± 25%. Stream gauging was most accurate in catchments with weirs (Emerald, Spuller, and Topaz) and in years when rating curves were based on constant-injection discharge measurements.

Errors introduced from analytical chemistry of snow samples had uncertainties of ± 2% to 7% for most solutes. Solute concentrations were higher in nonwinter precipitation, and for most solutes analytical errors were in the range of ± 1% to 5%, exceptions being sodium (± 10%) and sulfate (± 6%). In lake and stream samples, analytical errors were typically ± 2% to 6% but were ± 10% for ammonium owing to low or undetectable concentrations.

The uncertainties of VWM concentrations, fluxes, and yields were computed as the square root of the quadratic sum of the component errors (root-sum-square method; Sokal and Rohlf 1981). The estimated errors in VWM chemistry for each solute, from every year and at all catchments, are summarized by Melack et al. (1998). With the exception of 1990, errors in VWM concentrations were low and were mainly caused by analytical uncertainty. In the case of WY 1990, VWM rain values have standard errors of 15% to 30% owing to relatively large sampling error because rain was not sampled in the autumn of 1989. Solutes with concentrations at or below detection limits had the greatest uncertainty. Since solute fluxes were computed as the product of water volumes and VWM concentrations, the uncertainty in solute flux will derive from errors in both. The uncertainty in annual solute loading and solute yield was estimated by propagating the absolute errors in solute loading from snow, solute loading from nonwinter precipitation, and, in the case of yield, solute exports in outflow.

The solute balance component with the greatest degree of uncertainty was export. In most cases, errors were on the order of ± 20% to 30%, but in the case of ammonium and nitrate, larger errors were computed in some years. Solute exports had higher error for two reasons: (1) sampling errors were relatively high; and (2) volumetric errors for discharge were generally greater than for snow or rain. Accurate discharge measurements with weirs and daily chemistry samples during snowmelt were responsible for export uncertainties of less than ± 10% for most solutes. For nonwinter precipitation, errors were typically ± 10% to 25%. Uncertainty was lower for snow loadings since volumes were more accurately measured and spring snowstorms are comparatively rare in the Sierra Nevada. The uncertainties in solute yield were typically ± 5% to 20%.

Chloride yield can be used as an internal check on the accuracy of the solute balances because, with the exception of evapoconcentration, there are few biogeochemical or physical processes within the catchments that affect chloride concentration. Thus the ratio of chloride input to chloride export should be one, and chloride yield should approach zero. Excluding known and understood outliers, most chloride yields were in the range of ± 10 Eq ha^{-1}. In over half of the years chloride inputs and exports agreed within 10%, and in two-thirds of the cases the agreement was better than 20%. These findings suggest that chloride behaves as a conservative ion in Sierra catchments and that the chloride yields are consistent with the solute balance errors.

Loading of ions to the Emerald Lake catchment varied as a function of precipitation volume and solute concentrations in precipitation and the quantity of dry deposition of particles and gases (figure 28A). Atmospheric deposition at Emerald Lake is dominated by three constituents: hydrogen ion, ammonium, and nitrate. Hydrogen ion (105 moles ha^{-1}) and nitrate loading (123 moles ha^{-1}) are contributed mainly by dry deposition and likely occur in the form of gaseous HNO_3 deposition to catchment surfaces. Ammonium loading (91 moles ha^{-1}) is dominated by nonwinter wet deposition and likely results from atmospheric scavenging of NH_3 gas released by agriculture in the Central Valley of California. Loading of base cations (ranging from 8 moles ha^{-1} to 25 moles ha^{-1}) and chloride (31 moles ha^{-1}) is contributed mainly by winter snowfall. Sulfate loading (29 mole ha^{-1}) is somewhat evenly contributed by wet and dry deposition. There is no measurable ANC in atmospheric wet deposition in the Sierra Nevada.

Outflow export of ions depends on the concentration of solutes in catchment discharge and the volume of flow. The largest exports from the Emerald Lake watershed are the products of mineral weathering, including ANC (207 moles ha^{-1}), calcium (82 moles ha^{-1}), and sodium (107 moles ha^{-1}) (figure 28B). Export of the other base cations and acid anions ranged from 15 moles ha^{-1} to 25 moles ha^{-1}. Watershed processes strongly consumed hydrogen ion (7 moles ha^{-1}) and ammonium (3 moles ha^{-1}). Minor losses of gaseous N and S, mediated by soil microbes, can occur but were not measured.

Catchment yields reflect the internal supply and consumption of solutes (figure 28C). Hydrogen, ammonium, and nitrate ions are consumed by catchment processes, including titration of acids by base cations generated from weathering and biological assimilation of inorganic nitrogen, and, therefore, have negative yields. Base cations are generated by catchment weathering processes and soil cation exchange reactions and have positive yields. Chloride acts conservatively, with the inputs approximately balanced by outputs and with yields fluctuating between positive and negative values around zero. Sulfate concentrations in catchment streams are regulated by adsorption-desorption reactions in soils, and the fluctuation between positive and negative yields suggests that inputs and outputs of sulfate are in near-balance at Emerald Lake. The ratio of yield to loading (expressed as a percent) is an estimate of the efficiency of watershed uptake for ions with negative yields (hydrogen, ammonium, and nitrate) and watershed

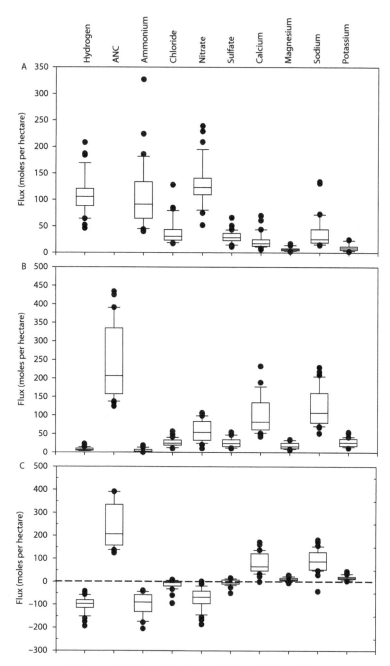

FIGURE 28. Box and whisker plot of annual solute balances for 1985–2016 at Emerald Lake. (A) Loading is the combined areal input of solutes from winter snow, nonwinter precipitation, and estimated dry deposition. (B) Outflow is the areal flux of solutes exiting the catchment via surface runoff. (C) Yield is computed as outflow minus loading; negative yields indicate net consumption of that solute by the catchment, while positive yield indicates net production.

production for ions with positive yields (base cations). For the 32-year record, the median yields, expressed as mole ha^{-1}, for H, NH$_4$, Cl, NO$_3$, SO$_4$, Ca, Mg, Na, K, and ANC were −97 (−92% of loading), −89 (−95%), −4 (−20%), −69 (−53%), −4 (−2%), 65 (+454%), 10 (+214%), 90 (+331%), 17 (+290%), and 207 (undefined % because of zero ANC loading), respectively. Despite substantial loading of hydrogen ion and ammonium, the Emerald Lake watershed consumes well over 90% of these inputs. In contrast, only about half of the nitrate deposition is assimilated by the catchment. Dry, barren soils in the Emerald Lake watershed have the capacity to generate high rates of nitric oxide evasion when wetted by precipitation, which may mean we are overestimating catchment nitrogen retention (Homyak et al. 2016). Both chloride and sulfate have relatively low consumption percentages, suggesting that they behave as conservative ions. And last, the high percentages for base cations indicates that the catchment generates two to five times more cations than are deposited via atmospheric deposition.

At Emerald Lake annual loading of base cations in snow is 23% of annual cation yield (i.e., production of cations by internal catchment processes such as weathering). Lower inputs of base cations in snow coupled with constant rates of acid deposition in dry deposition and rain may alter weathering rates in the Sierra Nevada as more cations will need to be weathered from bedrock to prevent declining alkalinity in surface waters. Interestingly, the yield of base cations from the Emerald Lake watershed was positive for all years except for 2015 (lowest outliers in figure 28C), which had the smallest snowpack on record for both Emerald Lake (see figure 15) and the overall Sierra Nevada. During 2015, we measured a net consumption of cations indicating that weathering produced fewer base cations than were deposited in atmospheric deposition. Thus acidification of surface waters in the Sierra Nevada might result in the future from a shift in the balance of inputs of acid and bases as snowfall declines.

Tokopah Basin

Mass balances of solutes for the larger Tokopah basin elucidate the same set of processes consuming and producing solutes in the Emerald Lake watershed, namely, mineral weathering, nitrogen assimilation, acid neutralization, and absorption-desorption of sulfate (figure 29). For the 24-year record, the median loads, expressed as 10^6 mole ha^{-1}, for H, NH$_4$, Cl, NO$_3$, SO$_4$, Ca, Mg, Na, and K were 73, 25, 125, 27, 11, 7, 19, and 10, respectively (figure 29A); these

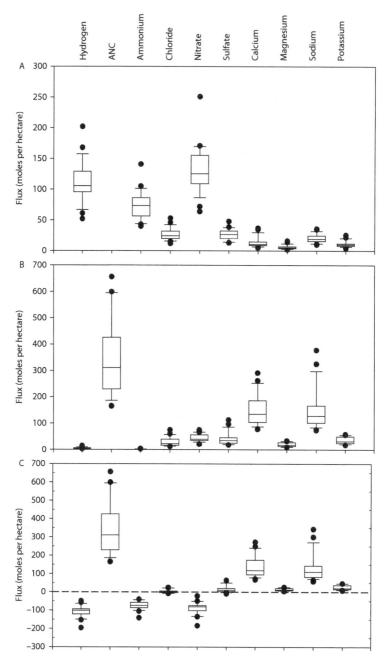

FIGURE 29. Box and whisker plot of annual solute balances for 1993–2016 for the Tokopah basin. (A) Loading is the combined areal input of solutes from winter snow, nonwinter precipitation, and estimated dry deposition. (B) Outflow is the areal flux of solutes exiting the catchment via surface runoff. (C) Yield is computed as outflow minus loading; negative yields indicate net consumption of that solute by the catchment, while positive yield indicates net production.

fluxes are similar to median loading in the longer Emerald Lake deposition record.

Differences between Emerald Lake and the Tokopah basin were detectable in the outflow export of solutes (figure 29B). While both catchments had low export of hydrogen (7 moles ha^{-1}) and ammonium (0.2 moles ha^{-1}), median areal fluxes of ANC (311 moles ha^{-1}), calcium (136 moles ha^{-1}), and sulfate (35 moles ha^{-1}) were about 50% higher in the Tokopah basin than in the Emerald Lake watershed. The Tokopah basin areal flux of sodium (129 moles ha^{-1}) was about 20% higher than at Emerald Lake.

Patterns for solute yield in the Tokopah basin closely followed those at Emerald Lake, with the exception of ANC, sulfate, and base cations (figure 29C). For the 24-year record, the median yield, expressed as moles ha^{-1}, and ratio of yield to loading (expressed as percent), for H, NH$_4$, Cl, NO$_3$, SO$_4$, Ca, Mg, Na, K, and ANC were −101 (−95% of loading), −73 (−100%), 2 (7%), −81 (−65%), 10 (+37%), 119 (+1082%), 12 (+171%), 110 (+579%), 21 (+210%), and 311 (undefined % because of zero ANC loading). In the Tokopah basin, median areal yield of calcium exceeded sodium yield, whereas at Emerald Lake, sodium yield was about 25% higher than calcium yield. A positive yield of sulfate in the Tokopah basin contrasts with the negative yield of sulfate from the Emerald Lake watershed. These observations suggest that the geology of the Tokopah basin is somewhat different from that at the Emerald Lake watershed and may contain small amounts of CaSO$_4$ minerals in remnants of the metasedimentary roof pendant of the Sierra Nevada.

Multi-Watershed Interannual Comparison

While the atmospheric deposition data presented in this multi-watershed comparison does not include dry deposition for the nonwinter period, comparisons among catchment solute balances are a valuable means of identifying important biogeochemical processes. By not accounting for dry deposition, estimates of catchment neutralization and assimilation of hydrogen, nitrate, sulfate, and ammonium are underestimated to various degrees. Similarly, catchment yields for base cations are overestimated.

Hydrogen ion: Annual loading of hydrogen ion ranged from 22.5 moles ha^{-1} to 128 moles ha^{-1} (table 5). Both the upper and lower extremes of wet deposition were measured in the Emerald Lake catchment; and wet hydrogen ion deposition at Emerald Lake varied by a factor of 6 during the period 1985 to 1994. Most of the wet hydrogen ion deposition in high-elevation

TABLE 5 Average of annual solute loading from 1985 through 1994 for the eight watersheds.

Solute	Mean	C.V.	Range
Hydrogen	51.9	0.48	22.5–128
Ammonium	50.8	0.47	20.8–141
Chloride	22.5	0.73	7.7–84.5
Nitrate	44.8	0.37	26.8–116
Sulfate	18.7	0.39	8.2–47.4
Calcium	14.0	0.39	8.0–33.7
Magnesium	3.7	0.38	1.8–8.7
Sodium	18.1	0.50	6.0–49.2
Potassium	7.7	0.54	2.6–22.6

NOTE: Loading is the sum of solute deposition from winter and nonwinter precipitation. Mean values for each major constituent (mole ha^{-1} yr^{-1}), coefficient of variation (C.V.), and the range of values are presented.
SOURCE: From Melack et al. 1998.

watersheds occurs during winter months. A comparison of winter and non-winter wet loading of hydrogen ion from 1990 through 1993 showed that 67% to 92% of the annual wet deposition was in winter snowfall (Melack, Sickman et al. 1997). The percentage of hydrogen ion deposition from winter loading is directly related to the quantity of snowfall; thus wet hydrogen ion deposition from nonwinter periods comprised a higher percentage of annual wet loading during years with low winter snowfall.

Average hydrogen ion yield was −45.2 moles ha^{-1} (range, −18.9 mole ha^{-1} to −111 mole ha^{-1}; table 6), which indicates that consumption of acidic inputs within all watersheds and lakes was high. Based on input-output budgets, the catchments best able to neutralize acidic inputs were Spuller, Ruby, and Crystal, and the mean percentage of hydrogen consumed in these catchments during WYs 1990 through 1994 was 94%. Spuller, Ruby, and Crystal Lake watersheds had significantly ($p < 0.05$) lower hydrogen ion yield than Lost Lake or the lakes in Sequoia National Park. The Lost Lake watershed consumed 82%, 57%, 62%, and 61% of hydrogen ion inputs for WYs 1990 through 1993, respectively.

Acid Neutralizing Capacity (ANC): Approximately 100 moles ha^{-1} to 500 moles ha^{-1} of ANC were exported annually (mean, 250 moles ha^{-1}; table 6). No ANC was measurable in precipitation. The Pear Lake watershed had significantly ($p < 0.05$) lower ANC yield (1990–93 mean, 138 mole ha^{-1}) than other catchments. From 1990 through 1992, ANC yields from the Spuller

TABLE 6   Average of annual solute yield from 1985 through 1994 (excluding 1988 and 1989) for the eight watersheds

Solute	Mean	C.V.	Range
Hydrogen	−45.0	0.48	−111–18.9
ANC	250	0.49	91.4–591
Ammonium	−50.5	0.46	−140–17.6
Chloride	−0.6	18.5	−35.2–25.3
Nitrate	−21.5	0.94	−62.5–24.8
Sulfate	8.5	1.62	−23.7–38.1
Calcium	85.5	0.57	21.3–204
Magnesium	12.8	0.75	3.1–49.8
Sodium	66.4	0.53	24.5–154
Potassium	18.1	0.68	−2.9–53.5
Silicate	252	0.57	66.1–642

NOTE: Yield is the difference between solute export and inputs and is expressed on an areal basis. Mean values for each major constituent (mole ha^{-1} yr^{-1}), coefficient of variation (C.V.), and the range of values are presented.
SOURCE: From Melack et al. 1998.

and Ruby watersheds were significantly ($p < 0.05$) higher than in other catchments; the largest ANC export was from the Spuller watershed (1990–94 mean, 304 moles ha^{-1}). Export of ANC was positively related to the quantity of runoff and was significantly ($p < 0.05$) greater in years with large snowpack, for example, 1993.

Chloride: The majority of wet chloride deposition occurred during the winter; nonwinter precipitation contributed less than 36% of annual wet chloride loading in 1990 through 1993. In large snowpack years, such as 1993, the annual contribution from winter snowfall represented an average of 94% of annual wet chloride loading at 15 stations in the Sierra Nevada (Melack, Sickman et al. 1997). The mean annual wet loading of chloride to the eight watersheds was 22.5 moles ha^{-1} (see table 6).

Chloride was not strongly retained or exported from the catchments, and there were no significant interannual differences in chloride yield for the period 1990 through 1993. With few exceptions, yields were in the range of ± 10 moles ha^{-1}, with the mean yield being −0.6 mole ha^{-1} (see table 6). Four outliers were noted in the chloride balances (WYs 1985 and 1986 at Emerald Lake, WY 1992 at Lost Lake, and WY 1993 at the upper Marble Fork). The Emerald Lake watershed anomalies reflect chloride contamination of snow samples (owing to residues from HCl cleaning of sample containers), and the other outliers are likely the result of errors in the water balances (Melack et al. 1998).

Sulfate: Wet deposition of sulfate during nonwinter periods accounted for the majority of wet annual loading in WYs 1990 through 1992 (mean for three years, 59%). During WY 1993, nonwinter precipitation comprised, on average, 4% of water inputs but contributed ~16% of wet annual sulfate loading. Nonwinter wet sulfate deposition is relatively large because of high concentrations of sulfate in summer rain (mean, 15.1 μEq L^{-1}). Mean wet sulfate loading was 18.7 moles ha^{-1} and ranged from a high of 47.5 moles ha^{-1} (Emerald Lake, WY 1987) to a low of 8.0 moles ha^{-1} (Tokopah basin, WY 1994). At Emerald Lake, high sulfate deposition in 1987 resulted from frequent, high-solute storms during spring snowmelt.

In 24 of 36 catchment-water years, sulfate yields were positive, and when yield was negative, few values were less than −7.5 moles ha^{-1}. The Ruby, Spuller, Marble Fork, and Lost watersheds had positive yields, while the Pear, Topaz, and Crystal watersheds had negative yields, except in years with large snowpacks. In 1990 through 1992, sulfate yield from the Ruby, Spuller, and Lost watersheds was significantly ($p < 0.05$) greater than from the Crystal, Pear, and Topaz watersheds. In the Emerald watershed, four years had positive yields (1986, 1991, 1993, and 1994), two had near-zero yields (1990 and 1992) and two had negative yield (1985 and 1987). The mean sulfate yield was 8.5 mole ha^{-1} (see table 6). The sulfate budgets indicate that there is a net export of sulfate from most watersheds in the Sierra Nevada during most years and that in wet years all Sierra catchments have positive yields of sulfate. A likely source of sulfate is mineral weathering of sulfur-bearing rocks; dry deposition may also be a source.

Base Cations: Calcium and sodium comprised the majority of base cation deposition to the catchments. Annual wet deposition rates of calcium ranged from 8.0 moles ha^{-1} to 38.5 moles ha^{-1}, and the mean was 14.0 moles ha^{-1} (see table 5). Mean loadings for magnesium and potassium were 3.7 moles ha^{-1} and 7.7 moles ha^{-1}, respectively. In dry to normal years (1990 through 1992), nonwinter precipitation comprised from 40% to 50% of the annual wet loading of base cations, and this percentage dropped to 7 to 12 in 1993.

High rates of base cation export were a conspicuous feature of the solute balances and were predominantly a result of mineral weathering. Calcium and sodium were the major cations exported; the mean and range of yields for calcium were 85.5 moles ha^{-1} and 21.3 moles ha^{-1} to 204 moles ha^{-1}, respectively, and for sodium,: 66.4 moles ha^{-1} and 24.5 moles ha^{-1} to 154 moles ha^{-1}, respectively (see table 6). Annual yields for magnesium and potassium ranged from −2.9 moles ha^{-1} to 99.5 moles ha^{-1} (means, 12.8 moles ha^{-1} and 18.2 moles ha^{-1}, respectively). Similar values for base cation yield were measured in the Emerald Lake watershed.

Nitrogen: Average annual wet ammonium deposition was significantly greater (p < 0.05) than wet nitrate deposition and ranged from 20.8 to 141 moles ha^{-1} with a mean value of 50.8 moles ha^{-1} (see table 5). Mean annual wet nitrate-loading varied from 26.8 moles ha^{-1} to 116 moles ha^{-1}, with a mean of 44.8 moles ha^{-1}. For both ions the highest loading rate was measured at Emerald Lake during WY 1987. Nonwinter precipitation is the major contributor of wet nitrogen loading during drought years. For the period 1990 through 1992, well over half of the annual wet deposition of ammonium and nitrate took place during the months of April through November. Similar patterns of nitrogen wet deposition were also seen in WYs 1985 through 1987 at Emerald Lake. Moreover, even in relatively wet years such as 1993, deposition from nonwinter precipitation was about 22% of wet annual loading.

Most of the nitrogen intercepted by high-elevation catchments in the Sierra Nevada is utilized by terrestrial and aquatic ecosystems. Utilization of ammonium from wet deposition averaged greater than 99% on an annual basis (see table 6), and it was one of the few solutes for which yields were not significantly (p < 0.05) higher during 1993. Negative yields for nitrate were measured in 32 of the 36 cases, indicating that the watersheds are usually a sink for nitrate. If dry deposition of nitrate during nonwinter periods was included in the budgets it is likely that all of these nitrate balances would have negative yield since the apparent positive yield of nitrate could be accounted for by additional dry deposition (see the more complete mass balances for Emerald Lake for all years that include dry deposition). At the Crystal and Lost Lake watersheds, nitrate utilization was in the range of 85% to 95%, but for the remaining catchments, uptake was typically on the order of 20% to 60%.

Nitrate yields, while predominantly negative, varied considerably both among catchments and from year to year. For example, nitrate yields at Emerald Lake varied from a low of −62 moles ha^{-1} during WY 1987 to a high of 25 moles ha^{-1} in 1993 when accounting for only wet deposition in the budget. Positive yield of nitrate (indicating a net source of nitrate in the catchment) was measured in only 4 of the 36 cases, and these years include 1986 and 1993 in the Emerald Lake watershed and 1993 in the Ruby and Spuller Lakes' basins. However, no net export of dissolved inorganic nitrogen

(DIN, i.e., nitrate plus ammonium) was measured at these four catchments despite the fact that mean nitrate yield during 1993 (−11 moles ha^{-1}) was significantly ($p < 0.05$) higher (2 to 3 times) than in all other years.

Relationships between Solutes, Snowfall, and Runoff

We found significant positive correlations ($p < 0.01$, Pearson product moment correlation) between loading of all measured constituents and snow water equivalence (SWE) measured in spring snowpack. SWE explained over 80% of the variability in winter deposition of hydrogen ion and sulfate and 75% and 84% of the variability in chloride and sodium, respectively. A relatively low percentage of the variation in winter loading of calcium, magnesium, and potassium (44%, 54%, and 23%, respectively) was explained by SWE.

Snowfall explained 59% of the variability in wet ammonium loading and 71% of the variation in wet nitrate loading. Linear functions fit to the nitrate-SWE relationship had a y-intercept of 6.1 Eq ha^{-1}, respectively, and was significantly ($p < 0.05$) greater than zero. We suggest that dry deposition of nitrate onto the snowpack is a possible explanation for the non-zero intercept. For comparison, during WY 1987, nitrogen loading from dry deposition in the Emerald Lake basin was estimated to be 5.2 Eq ha^{-1} during the winter based on measurements of nitrate deposition velocities determined at a NOAA dry deposition monitoring station in a forest clearing 7 km west of Emerald Lake (Williams et al. 1995). In Melack, Sickman et al. (1997) similar equations were fit for a larger set of Sierra stations, and the y-intercept for nitrate was 8.8 Eq ha^{-1}. For comparison, average annual dry deposition loading of nitrate at Emerald Lake was 64 Eq ha^{-1} over the entire period of record (see figure 25), suggesting that about 10% to 15% of annual dry atmospheric loading of nitrate occurs in the winter at Emerald Lake.

With the exception of nitrate and ammonium, there were significant ($p < 0.01$, Pearson product moment correlation) linear relationships between the export of solutes and runoff for the aggregate of all catchment during 1990–93 and separately for the Emerald Lake watershed using the entire data record (see figure 27C). Among all catchments, runoff explained 51% of the variability in annual ANC export. Differences in watershed geology, hydrologic flowpaths, soils, and perhaps vegetation account for the low predictive value of runoff in the aggregated data set. In contrast, in the Emerald Lake watershed runoff explained more than 98% of the variability in the annual ANC yield, and the relationship between runoff and ANC yield remained constant over a nearly fourfold range in runoff during 1990–93. A similarly high r^2 was observed for the relation between runoff and ANC yield when the other catchments were examined individually. The linearity of the runoff-export relationships indicates that all of the catchments produce ample ANC and are not at risk of acidification under present-day acid loadings. If the acid-neutralizing capacity of the watersheds were being approached, then we would expect some diminishment of ANC export at high runoff, that is, a decline in the rate of ANC yield per unit discharge. Using ANC export per unit runoff as an index of catchment sensitivity to acid deposition, Crystal and Ruby would be the least sensitive to acid deposition (slopes, 0.61 and 0.43, respectively) and Pear and Lost would be the most sensitive (slopes, 0.15 and 0.17, respectively).

R^2 values greater than 0.90 were found for runoff versus solute yields for sulfate, most of the base cations, and silicate in all of the catchments. However, r^2 values between runoff and the export of some solutes were not as high (i.e., chloride, 0.84; potassium, 0.75; hydrogen ion, 0.75). Though overall a weak, positive association between runoff and nitrate export was found, the intercatchment variability suggests that watershed processes exert a large degree of control on the export of nitrate, and nitrogen dynamics vary considerably from catchment to catchment. The significant ($p < 0.01$) ln-linear fit obtained for the relationship between runoff and nitrate export in the Emerald Lake watershed indicates that nitrate export per unit discharge declines as runoff increases.

GEOCHEMICAL PROCESSES INFLUENCING SOLUTE CONCENTRATIONS

We focus our discussion of geochemical processes influencing solutes in high-elevation watersheds on the Emerald watershed because of the wealth of information available for it. We first summarize aspects of the Emerald watershed pertinent to analysis of weathering and cation exchange processes. Sulfate regulation is then discussed.

The 120 ha Emerald watershed has 616 m of vertical relief and an outlet at 2,800 m. The sparse vegetation consists of scattered conifers (lodgepole and western white pine), low woody shrubs (*Salix* sp.), grasses, and sedges (Rundel, Neuman, and Rabenold 2009). Massive bedrock outcrops cover ~33%; unconsolidated sand, gravel, and talus cover ~23%; and the remainder is a mixture of rock and soil. Bedrock is mainly granite and granodiorite (Sisson and Moore 1984; Clow 1987; Shaw 1997). The granite (~75% of bedrock) is composed mostly of An_{24} plagioclase, potash feldspar, and quartz, with minor amounts of biotite and hornblende. The granodiorite (~15% of bedrock) is An_{24} and An_{42} plagioclase; alaskite and aplite (~10%) contain orthoclase, quartz, and plagioclase. Soils are poorly developed, shallow (~0.35 m), and acidic (Cryumbrepts and Cryorthents) (Huntington and Akeson 1986; Brown, Lund, and Lueking 1990). On well-drained slopes and ridges base cation exchange capacity is very low (4–5 keq ha^{-1}) and is moderate in wetter areas or near pines or shrubs (60–140 keq ha^{-1}); small areas of meadow soils (~1 ha, Cryofluvents and Cryaquepts) have higher exchange capacities (200–500 keq ha^{-1}). The primary clay minerals are hydroxy-interlayered vermiculite and kaolinite, with small amounts of gibbsite (Brown, Lund, and Lueking 1990; Brown and Lund 1991).

Williams, Brown, and Melack (1993) used a variation of the steady-state geochemical weathering model developed by Garrels and Mackenzie (1967) to calculate that all of the base cations in stream waters could be accounted for by back-reacting these solutes with secondary minerals to produce the bedrock minerals found in the basin, and their weathering model had a large discrepancy occurring during snowmelt runoff. It appears likely that the release of hydrogen ion during biological assimilation of ammonium could consume about 25% of the alkalinity produced by geochemical weathering and explain the discrepancy (Williams et al. 1995).

Based on two years of intensive study in two small catchments composed largely of exposed granodiorite (0.2 ha and 0.5 ha, with 25% and 10% soil coverage, respectively) close to the Emerald watershed, Williams et al. (2001) found that mineral weathering was the major source of solutes in runoff. They attributed the calcium export in excess of stoichiometric plagioclase weathering to dry atmospheric deposition, weathering of mafic minerals, and disseminated calcite. The mass balances for Emerald Lake show that atmospheric deposition of calcium can account for about 20% of the total watershed export of calcium, supporting an atmospheric source of excess calcium in stoichiometric calculations (see figure 28).

We examined further geochemical processes in the Emerald Lake watershed based on our extended time series of solute concentrations. The hypothesis that weathering is the proximate source of silica and base cations (A. Leydecker, J. Sickman, and J. Melack, unpublished analysis) is based on three assumptions: (1) most weathering occurs over the winter and is roughly

TABLE 7. Proportional mineral compositions of reactants and products used in a mass balance analysis (Garrels and Mackenzie 1967) of weathering at Emerald Lake, possible solutions (A and B, in mole ha⁻¹; derived using a system of linear equations from Finley and Drever 1997), and measured net annual export (mole ha⁻¹).

	Na	Mg	K	Ca	Si	Al	A[2]	B[3]
Reactants								
plagioclase[1]	0.3	0.0	0.0	0.7	2.7	1.3	130	
biotite[1]	0.00	1.15	0.92	0.00	2.79	1.29	16	16
calcite	0.0	0.0	0.0	1.0	0.0	0.0	12	
hornblende[1]	1.89	2.69	0.13	0.33	7.00	1.22	18	18
albite	1.0	0.0	0.0	1.0	3.0	1.0		91
anorthite	0.0	0.0	0.0	1.0	2.0	2.0		51
Products								
vermiculite[1]	0.0	1.5	0.0	0.0	3.0	4.0	35	35
kaolinite	0.0	0.0	0.0	0.0	2.0	2.0	36	47
Annual export	84	13	17	97	343	0		

NOTES:
[1] From Shaw 1998.
[2] Using the average catchment plagioclase composition, An_{30}.
[3] Solving separately for albite and anorthite results in An_{36} plagioclase.

proportional to the duration and depth of snow cover; (2) the efficiency with which weathering products are flushed from the catchment increases with snow depth; and (3) weathering during the falling hydrograph is proportional to discharge. While not definitive, due to variable mineral formulations and the large number of possible reactants and products, we performed a mass balance analysis based on the approach of Garrels and Mackenzie (1967) (table 7).

Vermiculite and kaolinite were selected as weathering products; biotite was chosen over K-feldspar as a source of potassium. In an experiment, Clow (1987) trickled snowmelt through columns of sand-sized particles of Emerald watershed granite and granodiorite and found that outflow concentrations were in general agreement with falling limb concentrations at Emerald Lake. The possibility of amorphous silicate dissolution as a contributing factor is not excluded as Williams et al. (2001) measured elevated silica concentrations, with no concurrent increase in sodium, in soil lysimeters at the end of snowmelt.

The Emerald watershed, as is common throughout the western United States (Stauffer 1990), has more calcium exported than can be accounted for by stoichiometric plagioclase weathering. The slope of the regression between concentrations of Ca and Na is 0.84 ($r^2 = 0.87$), analogous to an anorthite ratio of An_{44-46} vs. an estimated mean of An_{30} (Shaw 1997). Trace amounts of calcite and anorthite weathering occur in Emerald watershed rocks, so some combination of calcite, hornblende, and preferential anorthite weathering is a likely source of the excess calcium.

Although weathering is the ultimate source of alkalinity in surface waters, cation exchange was proposed to be the proximate mechanism for neutralizing acidic deposition in the Sierra Nevada (Williams, Brown, and Melack 1993). However, subsequent analyses showed that weathering is the proximate source of ANC and base cations in outflow from the Emerald Lake watershed. This conclusion is based on several lines of reasoning. First, the outflow flux of ANC and base cations vs. annual runoff is linear over a fivefold range in annual discharge; the cation exchange capacity of catchment soils is insufficient in deep snow years to produce this relationship (see figure 27C). Second, interannual variations in acidic deposition produce proportional changes in base cation and acidic anion export (e.g., ~invariant ratio of $Ca:SO_4$ in figure 27B), not the varying differences dictated by cation exchange equilibria. Third, outflow calcium concentrations are highly correlated with silica ($r = 0.94$), sodium ($r = 0.90$), and bicarbonate alkalinity ($r = 0.87$), indicating a common source. Fourth, interannual variations in runoff produce proportional changes in base cation and silica export (e.g., ~invariant ratio Ca:Na and Ca:silica in figure 27B), not varying differences dictated by cation exchange equilibria. That weathering is the proximate source of base cations and ANC in the Sierra Nevada means that despite having low ANC, its surface waters appear resistant to acidic atmospheric deposition.

During the weathering of silicate rocks, HCO_3^- ions released are derived from atmospheric CO_2, as illustrated by albite weathering (Suchet and Probst 1993):

$$2 NaAlSi_3O_8 + 2 CO_2 + 11 H_2O \rightarrow Al_2 Si_2 O_5 (OH)_4 + 2 HCO_3^- + 2 Na^+ + 4 H_4 SiO_4.$$

This general relationship can be combined with the yield of bicarbonate for watersheds to estimate the uptake of carbon dioxide by Sierra watersheds. Based on measurements in seven watersheds summarized in Melack et al. (1998), the area-weighted yield of bicarbonate (assuming the ANC is bicarbonate) is 216 moles HCO_3^- ha^{-1} y^{-1}. The area of the high-elevation Sierra

Nevada is approximately 20,000 km^2. Hence, the total weathering-based uptake of CO_2 is approximately 43,000 moles y^{-1}.

Possible mechanisms of sulfate regulation include pH- or discharge-dependent sulfate adsorption in soils. Williams et al. (2001) attributed variations in sulfate fluxes to elution from the snowpack and sulfate desorption. Williams and Melack (1997) found evidence that sulfate loading from precipitation and snowmelt was temporarily stored by catchment soils in two mixed conifer watersheds in Sequoia National Park and its release controlled by the extent of soil flushing. Clow, Mast, and Campbell (1996), in a study of surface water chemistry in the Upper Merced River basin (Yosemite National Park), reported that sulfate concentrations varied much less than other solutes and suggested sulfate levels are regulated, to some extent, by within-watershed processes such as sulfate adsorption onto the surfaces of Al and Fe sesquioxides. $\delta^{35}S$ was measured in the snowpack and Marble Fork discharge during June and July 1998 (J. O. Sickman, unpublished data). Radioactive ^{35}S is produced in the atmosphere and has a half-life of 87.4 days. Thus it can be used to quantify atmospheric sulfate in stream water. Older pools of sulfate in soil and groundwater will have little or no ^{35}S because it will have radioactively decayed. The measurements show that during the peak of snowmelt runoff, about half of the sulfate in the Marble Fork originated from melting snow and half was from production of sulfate in soils and vegetation.

The relative invariance of sulfate in the surface waters of the Sierra Nevada can be an advantage when observing lakes for recovery from acid deposition (Heard et al. 2014). The watershed processes controlling adsorption and desorption of sulfate appear to be sensitive to changes in in the rate of atmospheric deposition of sulfate (figure 30). Between 1983 and 2012, concentrations of sulfate decreased in rain, snow and total deposition at Emerald Lake (see figures 22–24). Similar downward trends in sulfate concentrations and loading were evident in wet deposition measured at nearby NADP monitoring stations in the Sierra Nevada (figure 30A). Declining sulfate loading appears to explain decreasing sulfate concentrations in Emerald Lake between the late 1980s and early 2000s and highlights the sensitivity of sulfur biogeochemistry to changes in both climate and atmospheric deposition. Importantly, sulfuric acid is a primary route for hydrogen ion loading to lakes of the Sierra Nevada. In a study of 41 lakes sampled in 2007 and 2008, Heard (2013) measured the concentration of spheroidal carbonaceous fly ash particles in shallow sediments and found a negative correlation with lake ANC (figure 30B). The concentration of fly-ash particles is a reliable estimate

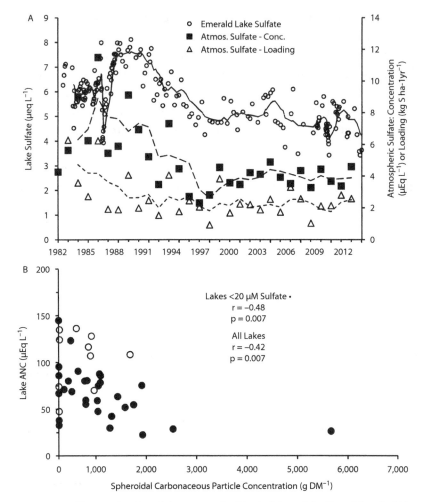

FIGURE 30. Temporal and spatial patterns of sulfate in lakes of the Sierra Nevada. (A) Time series of volume-weight mean chemistry for Emerald Lake from 1983 to 2012 in relation to sulfate concentrations and loading in atmospheric deposition. Lines denote 3-year moving averages. (B) Relationship between lake ANC and the concentration of spheroidal carbonaceous fly-ash particles (SCP) in surface sediments of 41 lakes sampled during 2007 and 2008 with varying sulfate concentrations (Sickman et al. 2013; Heard 2013). There is a significant negative correlation between ANC and SCP concentrations for lakes with < 20 µM sulfate and a weaker but significant correlation for all lakes.

of the rate of acid deposition in lake sediments (Rose et al. 2001). The significant negative correlation, especially in lakes without catchment sources of mineral sulfate (i.e., lakes < 20 μM sulfate), suggests that some lakes in the Sierra Nevada have their ANC depressed by current rates of acid deposition, and more improvements in air quality are needed to reduce deposition of sulfuric and nitrate acid below the critical load (Shaw et al. 2013).

USE OF MODELS TO UNDERSTAND AND PREDICT HYDROCHEMISTRY

Watershed models of varying complexity can be used to evaluate the relative importance of different processes and to predict changes in response to altered conditions, among other purposes. Models of relevance to watershed biogeochemistry are required to link hydrological and biogeochemical processes with appropriate spatial and temporal scales. As part of the California Air Resources Board (CARB) funded studies of potential impacts of acidic deposition on the Sierra Nevada, several such models were developed. Engle and Melack (1997) offer a critique of these models, only one of which has proved useful for the Sierra Nevada (alpine hydrological model, or AHM, Wolford, Bales, and Sorooshian 1996; Meixner and Bales 2003). An empirical modeling approach developed by Eshleman et al. (1995) was modified by Leydecker, Sickman, and Melack (1999) to estimate the effect of possible increases in acid loading to the Sierra Nevada. We discuss applications of these models to altered atmospheric deposition and climate in chapter 7.

As is evident from material presented in this and earlier chapters, snow cover influences biogeochemical processes. In this regard, hydrochemical modeling of the Emerald watershed by Meixner and Bales (2003) revealed that variations in duration of snow cover were more important to nitrogen mineralization and export than changes in atmospheric deposition of nitrogen. In another modeling study, Wolford and Bales (1996) compared the output of AHM generated for conditions in one year to its output when input data and parameters were modified to simulate changes in biological uptake of ammonium and nitrate, and routing 20% of runoff from rock surfaces directly to streams. If nitrate uptake was set to zero but ammonium uptake continued, large summer rain was predicted to cause depression in ANC and pH, presumably because hydrogen ions continued to be produced as ammonium was assimilated but were not being consumed by nitrate uptake. If 20% of rain or snowmelt runoff passed directly from rock surfaces to streams, ANC, pH, and calcium declined, presumably because of less time for weathering in soils.

NITROGEN DYNAMICS AND MASS BALANCES

Nitrogen supply and dynamics can control seasonal variation in nutrient limitation and productivity of terrestrial and aquatic ecosystems and is involved in the generation and consumption of hydrogen ion and ANC. Moderate rates of wet and dry deposition of N in the Sierra Nevada coupled with the oligotrophic nature of aquatic ecosystems make lakes and streams in the high elevations of the Sierra sensitive to nutrient inputs (Sickman. Melack and Clow 2003; Jassby et al. 1994). Nitrogen deposition in excess of uptake capacity has resulted in nitrogen saturation of terrestrial and aquatic ecosystems in forested watersheds in North America (Stoddard 1994; Aber et al. 1998), and evidence is growing that this process is occurring in high-elevation watersheds in the Colorado Front Range of the Rocky Mountains (Williams, Baron et al. 1996; Nanus et al. 2012).

Williams et al. (1995) synthesized results available in the mid-1980s for the Emerald Lake watershed and nearby sites to quantify sources, sinks, and transformations of nitrogen. General features of their analysis indicate that deposition of snow and rain accounted for most inputs of inorganic nitrogen; changes in deposition by throughfall were insignificant; and nitrogen fixation, though not measured, is likely low. Soils represented the largest reservoir of nitrogen, with vegetation and litter adding about 13% of that in soils and soil water; groundwater and the seasonal snowpack account for less than 1% of nitrogen storage in the watershed. Net mineralization of soil organic matter and biological uptake dominated fluxes to and from the soil water. Most export from the basin occurred as nitrate and organic nitrogen via stream discharge, while outputs as ammonium, N_2 or N_2O emissions, were quite small. Overall, inputs exceeded outputs, with the watershed accumulating about a quarter of the atmospheric inputs.

Interannual variations in mass balances of nitrogen for the Emerald Lake watershed (WYs 1985 through 1998) and six other Sierra basins (WYs 1990 through 1993) are reported by Sickman, Leydecker, and Melack (2001). Nitrogen retention was found to be high in most regions of the Sierra Nevada.

Solute balances indicate that catchments were net sinks for nitrate in most years and for ammonium plus nitrate in all years (see figure 28). In the Emerald Lake watershed, nitrogen mass balances, patterns of stream chemistry, and isotopic analyses indicate that nitrate export is the net result of nitrate flushed from soils and of snowpack nitrate that escapes biological uptake (Sickman, Leydecker et al. 2003). Stable isotope analysis of nitrate provides an important tool to trace sources and transformations of nitrate in catchments. In the Sierra Nevada, the $\delta^{18}O$ of atmospheric nitrate tends to be highly enriched (+60 to 100 permil) and the $\delta^{15}N$ of nitrate slightly depleted (−5 to 0 permil) (Homyak et al. 2014). In contrast, nitrate from catchment sources such as soils tends to have $\delta^{18}O$ and $\delta^{15}N$ values in the range of +10 to −5 and −15 to +5, respectively (Homyak et al. 2014). Thus measurements of $\delta^{18}O$ of nitrate can help determine the relative contributions of atmospheric and soil N to lakes and streams (Sickman, Leydecker et al. 2003).

Inputs of nitrate from the atmosphere to lakes can occur during snowmelt (Sickman, Leydecker et al. 2003) and as a result of rain falling directly on lake surfaces and dry deposition being washed off of catchment surfaces into streams. In 1994, following a dry summer, a series of large rain and snow events occurred during late September and early October. Peak nitrate concentration in the upper Marble Fork was 55 μEq L^{-1} 24 hours after the first rain event, while nitrate in the precipitation was less than 30 μEq L^{-1}. High nitrate levels immediately after the rainstorm are likely the result of dry deposition being washed off catchment surfaces and into streams. Data from 2011 (a large snowpack year) at Emerald Lake show the effects of direct rainfall on nitrate pools in lakes (figure 31A). During 2011, ice cover lasted into July and discharge remained above 10,000 m^3 day^{-1} through August so that early rain had no effect on lake nitrate. In September as discharge fell, a series of rain events caused the $\delta^{18}O$ and $\delta^{15}N$ of nitrate to change in lake surface waters, moving toward an atmospheric endmember with higher $\delta^{18}O$ and lower $\delta^{15}N$. Similarly, lake samples collected during the later summer of 2012 in Sequoia and Kings Canyon National Parks show a wide range in $\delta^{18}O$ of nitrate (figure 31B). About two-thirds of the lakes had nitrate that plotted fully within the isotope space defined by soils; however, about one-third of the lakes had $\delta^{18}O$ values intermediate between the soil and atmospheric endmembers, with two lakes having $\delta^{18}O$ of nitrate above 50 permil, indicating substantial atmospheric nitrate was present in the water column. These data demonstrate that while snowmelt runoff dominates the supply of nitrogen to Sierra Nevada lakes, direct precipitation, including dry deposition, can

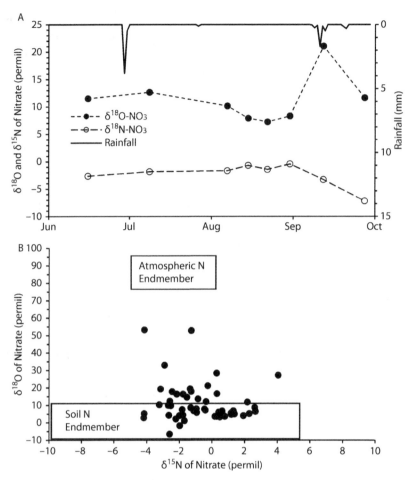

FIGURE 31. Patterns of isotopic variation in nitrate in lakes of the Sierra Nevada. (A) Time series of nitrate isotopes in Emerald Lake during 2011 relative to the timing of rain events. (B) Two-dimensional isotope space for isotopes of nitrate in 50 lakes sampled across the southern and central Sierra Nevada during 2007 and 2008 (Heard 2013).

be an important nutrient source when lakes and streams become hydrologically disconnected from catchments in the autumn as runoff declines and the lakes are thermally stratified.

At Emerald Lake, annual export of nitrogen varied by a factor of 8 from 1985 to 2016 (see figure 28) and was a linear function of runoff (r^2 = 0.89 and 0.74 for dissolved inorganic nitrogen and dissolved organic nitrogen, respectively) (Sickman, Leydecker, and Melack 2001). Nitrogen yield increased faster than increases in runoff, indicating ecosystem processes enhanced

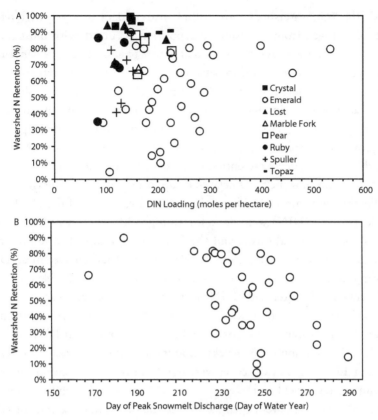

FIGURE 32. Relationships between watershed N retention (expressed as a percentage of loading) and loading (A) and day of peak snowmelt discharge (B). Panel A presents results from Emerald Lake along with data from 7 other catchments with multiple years of measurements.

nitrogen losses during years with high runoff and retarded losses during dry years. The timing of snowmelt runoff had a strong effect on inorganic nitrogen dynamics: nitrate pulses were greater and watershed nitrogen retention was lower in years with deep, late melting snowpack (figure 32) when the residence time of water in many small Sierra lakes can be on the order of days. Labile nitrogen pools in subnivean soils are augmented and plant uptake is reduced during years with high snowfall amounts, causing an imbalance between nitrogen supply and terrestrial nitrogen demand. Furthermore, during large snowmelt years, lake temperatures are low, and lakes and streams do not become ice-free until July, thereby reducing aquatic N demand.

Of the eight watersheds with multiple years of data, relatively high retention of nitrogen from atmospheric deposition occurred, with most catchments in most years having retention of > 70% of atmospheric N deposition (see figure 32). No relationship between the level of nitrogen retention and the quantity of inorganic nitrogen loading to the catchments was observed; most catchments maintained high nitrogen retention despite high variability in nitrogen loading. However, in the case of Emerald Lake, the longer data set provides examples of years when N retention dropped below 50% and to less than 10%. These low retention years tended to be ones when snowpacks were deep and snowmelt was delayed by up to a month from average.

Further comparative analyses of yields and retention of dissolved inorganic nitrogen (DIN) were done for 28 high-elevation watersheds in the Sierra Nevada of California and the Rocky Mountains of Wyoming and Colorado (Sickman, Melack, and Stoddard 2002). They reported mean DIN loading to Rocky Mountain watersheds of 3.6 kg ha^{-1} yr^{-1}, double the mean in Sierra Nevada watersheds. DIN yield in the Sierra Nevada was about 60% of that measured in the Rocky Mountains. Net inorganic N retention in Sierra Nevada catchments represented about 55% of annual DIN loading, while DIN retention in the Rocky Mountain catchments was 72% of DIN loading. Elevation and soil cover were significantly (p < 0.1) correlated with catchment DIN yield and retention in the Sierra Nevada; log-linear regressions based on soil cover explained 82% of the variation in catchment DIN retention. In the Rocky Mountains, 90% of the variation in DIN retention was explained by log-linear models based on soil cover.

One pronounced feature of nitrogen dynamics in Sierra watersheds is the release of nitrogen from the snowpack and its export from the catchments during snowmelt (see figure 26B). Two possible mechanisms have been invoked to explain the snowmelt pulse of nitrate. In one, atmospherically derived nitrate is preferentially eluted from the snowpack and transported to streams during the first fractions of snowmelt (Williams, Brown, and Melack 1993; Williams et al. 1995). As a refinement, Williams et al. (1995) postulated that ammonium is rapidly nitrified in the snowpack and soils and that this represents a significant percentage of the nitrate pulse in streams during snowmelt. In a second mechanism, nitrate derived from microbial activity in soil is flushed into streams by snowmelt waters. This mechanism was examined in the Loch Vale watershed in the Rocky Mountains and in the Emerald Lake watershed (Campbell et al. 2002; Sickman, Leydecker et al. 2003).

Using isotopic data, soil water measurements from lysimeters, and watershed mass balances, Sickman, Leydecker et al. (2003) concluded that 50% to 70% of nitrate export from the Emerald Lake watershed during snowmelt runoff was derived from soils, with the contribution from soil water greatest at the onset of snowmelt. Nitrate concentrations rose to a maximum a few weeks before peak runoff, suggesting sequestration and release of nitrate from microbial pools in soils and talus.

Once insulated by snow, the Sierra soils remain just above 0°C for the duration of the winter. To better understand the processing of nitrogen in cold, under-snow soils, Miller et al. (2007) conducted laboratory incubations at temperatures of -2°C, 0°C, and 5°C and water-holding capacities of 50% and 90% with soils from the Emerald and Topaz watersheds. They found that dry meadow soils mineralized nitrogen at -2°C and that wet meadow soils switched from net N consumption at -2°C to net nitrogen mineralization at temperatures $\geq$ zero. As net nitrate production decreased, ammonium continued to accumulate in soils.

To investigate seasonality in nitrogen biogeochemistry in Sierra soils, Miller et al. (2009) conducted year-round field incubations, gas flux measurements, and ^{15}N tracer additions in late autumn, early spring, and summer. From the start of snowmelt to early summer, coincident with increased microbial nitrogen, high ^{15}N uptake and turnover occurred. As snowmelt progressed, net nitrogen mineralization and net nitrification in wet meadow soils and net production of ^{15}N labeled ammonium in wet and dry meadow soils were detected. Net nitrogen mineralization and net nitrification rates were negligible in dry meadow soils on all but one sampling date. Release of N_2O in late spring and early summer, when soils were wet and above zero, represented roughly 3% of annual nitrogen inputs (Miller et al. 2009). These experiments show that ample labile nitrogen is present in soils at the onset of snowmelt to explain the stream nitrate pulse observed later in snowmelt.

Soil water collected from five sites in the Emerald basin had low concentrations of nitrate and ammonium prior to and after snowmelt (Williams et al. 1995). Low nitrate concentrations in soil waters prior to melt could be the result of low free-water content in these soils. In contrast, concentrations of nitrate in shallow soil solutions on a ridge in the Emerald basin exceeded 70 μEq L^{-1} for two consecutive weeks during peak snowmelt runoff. These nitrate levels are similar to those found in the lake and outflow at Topaz watershed during baseflow periods in autumn and winter. In the Emerald watershed the highest nitrate and ammonium concentrations measured in

TABLE 8. Landscape characteristics for fifteen high-elevation watersheds in the Sierra Nevada. Soil cover is expressed as a percentage of total catchment area, and mean slope is in degrees. Elevations are measured at the catchment outlet.

Catchment	Elev. (m)	Area (ha)	Soil Cover (%)	Mean Slope	Lake Area (ha)	Lake Vol. (10^3 m^3)	Water Year Record
Crystal	2,951	135	53	21°	5.0	324	1990–1993
Emerald	2,800	120	22	29°	2.7	162	1985–2008
Lost	2,475	25	36	14°	0.7	12.5	1990–1993
Marble Fork	2,621	1908	40	18°	–	–	1993–1999
Pear	2,904	142	22	24°	8.0	591	1990–1993
Ruby	3,390	441	18	27°	12.6	2,080	1990–1994
Spuller	3,131	97	33	22°	2.2	34.7	1990–1994
Topaz	3,218	165	41	10°	5.2	76.9	1990–1999
High	3,603	15	5	17°	1.0	17	1993–1994
Low	3,444	225	8	26°	0.2	1.1	1993–1994
M1	3,078	106	20	18°	0.6	7.0	1993–1994
M2	3,188	90	18	11°	0.5	5.2	1993–1994
M3	3,249	67	10	11°	0.5	5.2	1993–1994
Mills	3,554	177	6	26°	2.4	72	1993–1994
Treasure	3,420	175	10	29°	2.7	88	1993–1994

snow lysimeters were 28 μEq L⁻¹. Low nitrate concentrations in soil water following snowmelt is likely the result of plant uptake and storage within the terrestrial watershed.

The evidence that nitrate concentrations can vary as a function of timing of snowmelt, soil conditions, and location within a watershed suggests that varying spatiotemporal patterns of snowmelt may result in differences in stream nitrate concentrations. To investigate this possibility, Perrot et al. (2014) used twelve years of hydrometeorological and hydrochemical observations combined with a spatially distributed snowmelt model to link nitrate concentrations to spatially distributed snowmelt in the Tokopah basin. They determined the relationship between new snowmelt areas (NSA; based on modeled pixel-by-pixel melt duration) and stream nitrate. In the Tokopah basin, relationships between NSA indices and stream nitrate indicated that nitrate concentrations tended to increase when NSA was expanding and to decline once the whole watershed was melting; this patterns suggests that labile soil nitrate can be quickly exhausted as the rate of meltwater flushing increases during snowmelt. As much as 76% of the variability in stream nitrate was explained by the accumulated depth of snowmelt, which demonstrates the importance of hydrological connectivity between hill slopes and streams in the Tokopah basin.

To examine further relationships between hydrological processes and nitrate export, we computed characteristic flushing times for nitrate from fifteen Sierra Nevada watersheds (table 8) and compared these times to the relative amount of nitrate exported. Using an approach similar to that used by Creed and Band (1998), we computed catchment-specific export coefficients by regressing annual DIN export (dependent variable) versus annual discharge (independent variable) for each of 15 watersheds (figure 33). An exponential decay model was fitted to the decline in nitrate concentrations during snowmelt runoff during each year at each catchment:

$$N = N_i e^{-kt}$$

where:
N_i = the nitrate concentration at the peak of the snowmelt pulse,
t = time in days, and
k = the exponential decay coefficient.

Time constants ($t_c = 1/k$), the time required for peak concentrations to decrease by 37%, were calculated for each catchment year; values for multiple years were averaged to yield a single catchment-specific value. The export coefficient residuals (i.e., catchment-specific minus mean catchment export behavior) were regressed against the time constants and the regression residuals recorded. These residuals indicate whether the quantity of nitrate exported from a catchment is proportional to the amount of time that nitrate is being flushed from soils. The catchments in this example included the seven lake watersheds with multiple years of data, plus several smaller catchments located at high elevation with little soil cover, for example, High, Low, M1, M2, M3, Mills, and Treasure (figure 33C). Catchment soil cover in these small, highest-elevation catchments was less than 10%, while the large lake watersheds had between 20% and over 50% soil cover.

Catchment DIN flushing coefficients in these fifteen catchments varied by a factor of about 23 (8.0×10^{-5} kg-N ha^{-1} mm^{-1} at Crystal Lake watershed to 182×10^{-5} kg-N ha^{-1} mm^{-1} at High Lake watershed) with an average flushing coefficient for the fifteen study sites of 77×10^{-5} kg-N ha^{-1} mm^{-1} (figure 33B). Flushing residuals (calculated as a percent of the overall mean) ranged from -90% at Crystal Lake to $+138\%$ at High Lake. Catchment time constants ranged from 13 days at Lost Lake to 101 days at Ruby Lake; the mean catchment time constant was 26 ± 5 days. The unusually long time constant at Ruby Lake may be a result of appreciable groundwater input into the lake's relatively large area and volume. A plot of export residuals vs. nitrate flushing time constants (figure 33A), revealed two relationships: (1) catchments with below average DIN export had a positive correlation between export and flushing time; and (2) catchments with above average export had a negative correlation between export and flushing time. Catchments with little soil cover had above average DIN export, but the quantity of nitrate exported from a catchment was not proportional to the amount of time that nitrate was flushed from soils (figure 33C). While we do not have isotope data on nitrate sources for these catchments, it is likely that preferential elution of nitrate from the snowpack produced brief but intense periods of nitrate delivery to streams in catchments at the highest elevations with poorly developed soils, possibly inducing episodic acidification of lakes and streams (Stoddard and Sickman 2002). In catchments with soil cover over 20%, processes within soils (e.g., mineralization and nitrification) produced labile nitrate in soils during the initial stages of snowmelt and then infiltrating snowmelt displaced the nitrate over several weeks to months, resulting in a positive relationship between flushing time and export residuals but overall below average areal yield of nitrate.

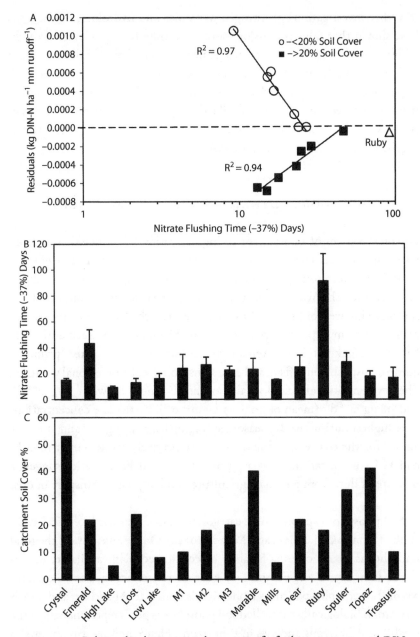

FIGURE 33. Relationship between catchment-specific flushing constants and DIN export residuals (i.e., the difference between catchment-specific regression coefficients and the overall mean) for the 15 watersheds listed in table 8. (A) Relationship between the characteristic flushing behavior of the catchment (x-axis) vs. that catchment's deviation from average nitrogen flushing response (y-axis). (B) Characteristic flushing behavior for the individual catchments. (C) Soil-covered area in the catchments.

Other investigations of nitrogen dynamics and hydrological flowpaths have shown that talus is a major source of nitrate in high-elevation catchments and that hydrologic transit times associated with talus are rapid (Williams, Brady, and Willison 1997; Bieber et al. 1998; Campbell et al. 2000). Short hydrologic residence times in rock- and talus-dominated catchments combined with the lack of soil to biologically mediate nitrate concentrations may explain the brief and intense nitrate pulse causing the interesting dichotomy illustrated in figure 33A.

PHOSPHORUS DYNAMICS AND MASS BALANCES

Potential sources of phosphorus to high-elevation lakes include flux from lake sediments (see chapter 6; see also Homyak et al. 2014b), release of labile phosphorus from soils, and aeolian transport of phosphorus-bearing dusts, aerosols, and ash. Atmospheric particulate matter sampling conducted at a station 10 km west of Emerald Lake deployed a stacked filter unit and micro-orifice uniform deposit impactor (Vicars, Sickman, and, Ziemann 2010; Vicars and Sickman 2011). Aerosol concentrations were elevated, primarily due to transport from offsite and emissions from local and regional wildfires. The dry depositional flux of total phosphorus ranged between 7 $\mu g\ m^{-2}\ d^{-1}$ and 118 $\mu g\ m^{-2}\ d^{-1}$ (mean of 40 ± 27 $\mu g\ m^{-2}\ d^{-1}$). Particulate concentrations were highest during the dry season, averaging 8.8 ± 3.7 $\mu g\ m^{-3}$ and 11.1 ± 7.5 $\mu g\ m^{-3}$ for the coarse and fine fractions, respectively, while winter months had PM concentrations of < 1 $\mu g\ m^{-3}$. Fe:Al and Fe:Ca ratios suggest a mixture of dust from regional agriculture and long-range transport of dust from Asia.

Soil phosphorus pools can be categorized based on their biological availability (Homyak, Sickman, and Melack 2014a). On average, 14% of the total phosphorus in A-horizons is labile or considered plant available, 62% is mostly bound to Al and Fe, and 24% is considered refractory. For B-horizons, 9% of the total phosphorus is labile, 61% is bound to Al and Fe, and 30% is refractory. Biologically controlled soil phosphorus pools represent 62% of the total-P in A-horizons and 53% in B-horizons. Hence phosphorus appears to not be in short supply in soils within the Emerald watershed. Phosphorus concentrations in microbial biomass are highest during winter when soils are snow covered, and concentrations decline over time into the autumn. This pattern indicates the importance of microbial P immobilization in subnivean soils and highlights a potential mechanism for soil phosphorus retention that may be sensitive to climate change.

Based on a survey of fifty Sierra lakes, stepwise multiple linear regressions did not indicate that sediment phosphorus chemistry was an important predictor of soluble reactive phosphorus or total dissolved phosphorus in lake water (Homyak, Sickman, and Melack 2014b). In contrast, these regressions support watershed sources for phosphorus since positive relationships were detected between lake water sodium concentrations and labile phosphorus and between silica, a weathering product, and total dissolved phosphorus.

In summary, watershed inputs of phosphorus appear to be the main source to high-elevation Sierra lakes. In the case of the Emerald Lake watershed, direct aeolian inputs into the lake are approximately 0.3 kg yr^{-1} (Vicars, Sickman, and Ziemann 2010), and the catchment exports are 8.8 kg yr^{-1}, which includes atmospheric deposition and phosphorus mobilized by soil processes (Homyak, Sickman, and Melack 2014a).

SIX

Limnology

Abstract. Studies of high-elevation Sierra lakes include extensive sampling of numerous lakes, intensive measurements, experiments, and long-term monitoring of selected lakes, with a focus on Emerald Lake and its watershed. Physical processes that form the density structure of lakes, chemical conditions and lacustrine metabolism, and eutrophication and acidification are discussed. Emerald Lake, a representative Sierra lake, is chemically very dilute and weakly buffered; calcium and bicarbonate are the two major ions. Episodic ANC depletion during snowmelt occurs mainly through dilution by low ANC snowmelt rather than by addition of strong acids in meltwaters. Experimental acidification in stream channels and in-lake mesocosms detected negative effects on some biota. Dissolved oxygen in surface waters generally remained above 5 mg L^{-1} in all lakes; however, hypoxia or anoxia may develop under ice in some years or in deeper lakes. Dissolved organic matter concentrations in Sierra lakes tend to be low and vary in source and bioavailability seasonally or in relation to landscape differences in vegetation cover. High-elevation lakes are likely to be near metabolic balance because terrestrial organic matter inputs are low, and low nutrient concentrations limit aquatic primary production. Benthic rates of production may account for a majority of whole-lake production at certain times of year. The major drivers of ecological change in Sierra lakes during the twentieth century have been trout introductions (with negative effects on native amphibians and invertebrate species) and acid deposition (with widespread reductions in lake ANC and changes in diatom taxa).

Key Words. physical processes, eutrophication, acidification, metabolism, calcium, bicarbonate, dissolved oxygen, primary production, trout, amphibians

HIGH-ELEVATION LAKES ARE OF INTEREST for several reasons. They are globally distributed and responsive to short- and long-term environmental change (Moser et al. 2019). Weathering rates in mountain systems are usually low, a key factor contributing to their dilute waters, low nutrient inputs, high water clarity, and susceptibility to altered atmospheric inputs. Long periods of snow and ice cover and large seasonal differences make them sensitive to changes in climate. Steep elevational gradients in mountains, coupled with a high degree of spatial heterogeneity in landscape and lake morphometric features, provide a framework for exploring how abiotic factors regulating ecosystem function are mediated by regional and local factors. Comparatively simple food webs make high-elevation lakes excellent systems in which to link ecosystem function and structure and examine the comparative roles of top-down and bottom-up processes. Finally, their locations can lead to them being less disturbed by anthropogenic effects. While their remoteness offers advantages, it is a logistical obstacle to their study.

One of the requirements for understanding high-elevation Sierra lakes is sampling year-round. Instruments required to monitor environmental conditions must withstand deep snow and severe weather. Nonautomated sampling is challenging because almost all the lakes are accessible only by foot. While some lakes can be reached after hiking kilometers of mountain trails that traverse steep terrain, most lakes are located off trail, requiring cross-country navigation. Access is further complicated in the winter, when the Sierra Nevada is blanketed by several meters of snow for more than half the year. For example, Emerald Lake is located 8 km and about 600 m above the trailhead in Sequoia National Park and is accessible only by foot or ski.

Early aquatic ecological studies in the Sierra Nevada were motivated by interest in introduced fishes. Work done by personnel at the Convict Creek Experiment Station, established in the mid-1930s in the eastern Sierra, included limnological studies of the lakes in the Convict Creek basin by Reimers, Maciolek, and Pister (1955). This station, now part of the University of California's Natural Reserve System and called the Sierra Nevada Aquatic Research Laboratory (SNARL) continues to support studies of Sierra lakes and streams. The first year-round study of a high-elevation Sierra lake, reported by Stoddard (1987), was done from SNARL. Concerns about potential impacts of acidic atmospheric deposition on high-elevation lakes resulted in considerable research in the 1980s and 1990s (e.g., Melack and Stoddard 1991). These studies began with extensive sampling of high-elevation lakes

(Melack, Stoddard, and Ochs 1985) and evolved into intensive measurements, experiments, and long-term monitoring of selected lakes, with a focus on Emerald Lake and its watershed. While selected aspects of these studies are available in the published academic literature, much remains in reports to the California Air Resources Board (summarized in Engle and Melack 1997) and as unpublished data obtained as part of other work. In this chapter we blend and synthesize these data and data from numerous other publications.

Melack and Schladow (2016), in their overview of lakes in California, include an introduction to basic limnological terminology and brief information about high-elevation lakes of the Sierra and two large lakes on its eastern flank (Lake Tahoe and Mono Lake). Here we begin with the physical processes that form the density structure of lakes and then consider chemical conditions and lacustrine metabolism, followed by eutrophication and acidification, as possible perturbations to the lakes' ecology.

PHYSICAL PROCESSES

Water temperature is perhaps the single most important abiotic factor governing aquatic ecosystems. Temperature affects the metabolic rates of organisms and governs the saturation of dissolved gases such as oxygen and carbon dioxide. Variability in temperature through the water column is a large determinant of water density, which structures physical dynamics. Most Sierra lakes have seasonal patterns in temperature: lakes are thermally stratified under ice, vertically mix to some extent after ice cover melts, restratify in the spring, and remain stratified throughout the summer, mixing thoroughly for a period of weeks again in the autumn (figure 34). A pattern of complete mixing of the water column twice annually is a fundamental feature of most, though not all, Sierra lakes. Lakes that are shallow, or located in basins that are especially windy, may mix multiple times throughout the summer, and some lakes may mix incompletely during snowmelt before stratifying in the spring. Regardless of the annual mixing frequency, during the winter Sierra lakes become inversely stratified under ice. Temperatures near the bottom reach a maximum of ~4°C and become colder with decreasing depth, often approaching 0°C in proximity to the ice.

The ice cover found on Sierra lakes differs from most other temperate and arctic lakes. Rather than ice forming only through freezing of lake water, Sierra lakes have a series of ice lenses sandwiched between layers of slush that

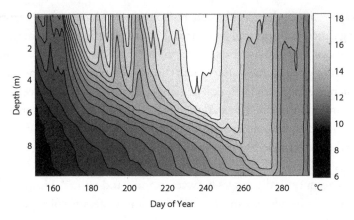

FIGURE 34. Emerald Lake, ice-free period (2016): Time-depth diagram of isotherms. Emerald Lake is dimictic, stratifying during the summer and mixing during the spring and autumn.

form when snowfalls depress layers of ice. With successive winter storms, this amalgam of hard ice and slush grows, commonly reaching between 1 m and 3 m in thickness, and blocks direct solar heating of the water column. In shallow lakes, such as Spuller and Topaz, the ice layers can occupy about 90% of the lake volume and may limit overwinter habitat for aquatic organisms such as fish or amphibians. The thickest ice cover measured on Emerald Lake occurred during the winter of 1986 when, due to abundant snowfall and a large avalanche, the lake had 6 m of slush and ice; over 75% of the lake's volume was occupied by this material.

During snowmelt, lakes are flushed and have short residence times given the comparatively small volume of most lakes with respect to the volume of inflowing meltwaters. Nelson, Sadro, and Melack (2009) calculated theoretical hydraulic residence times from 3 to 68 days during the summer of 2005 for Emerald Lake as the summed number of days of discharge required to exceed the lake volume. Sadro, Melack, and MacIntyre (2011a) calculated a residence time at ice-off in 2007, a year with below average snowpack, as ~30 days with discharge below 20 L s^{-1}. At ice-off in 2008 discharge was above 100 L s^{-1} and residence time was ~10 d. Expressed as a percentage of the lake volume, daily flushing rates during spring snowmelt in Emerald Lake span an order of magnitude, ranging from 0.8% to 10.9 % in relation to variation in winter snowpack (Sadro et al. 2018). While the relatively small volume of most Sierra lakes (see table 1) suggests that in years with average snowfall the entire lake volume is replaced, complexities associated with the flowpath of meltwaters may make actual water residence times longer than theoretical flushing times, as illustrated by Cortes, MacIntyre, and Sadro (2017).

By June or July of most years, ice has melted and the majority of snowmelt is complete (see chapter 4), after which the temperature of the upper layers of lakes increases (figure 34). Peak lake temperatures in the upper teens or low 20s °C are often reached by early August, though maximum temperature tends to increase with decreasing elevation and shorter periods of ice cover. Sierra lakes are sufficiently clear that direct solar heating at depth can occur. The attenuation coefficient of photosynthetically active radiation in lakes above 2,500 m typically ranges between 0.2 m^{-1} and 0.5 m^{-1}. However, lakes or ponds near meadows can have less clear waters because of inputs of dissolved organic matter.

The thermocline typically develops during or soon after ice-off and strengthens through midsummer before beginning to erode. Schmidt stability, which reflects the amount of energy required to overcome resistance to mixing in the water column, reaches maxima from late July through early September (Sadro, Melack, and MacIntyre 2011a). However, nighttime air temperatures can be sufficiently low during the summer to promote convective mixing of the upper water column. On cloud-free days a diel thermocline develops within the top 2 m of the water column during the day and is eroded by nighttime cooling. The seasonal thermocline is gradually eroded through a combination of convective mixing and a reduction in the strength of stratification as deeper waters warm through the summer. By mid-September in most years, nighttime air temperatures are sufficiently cool to drive complete convective mixing of the water column on a nightly basis, although daytime warming and near-surface thermoclines can form until lake ice forms. The rate at which Sierra lakes cool in the autumn will depend on a number of terms in their energy budget and on their morphometric aspects. Small lakes are capable of cooling rapidly; by autumn Emerald Lake is capable of losing over 1°C per day.

Water temperature in Sierra lakes can vary considerably between years with large and small snowpacks. Differences in water temperature between years are the result of net changes in a lake's heat budget. Computations of surface energy fluxes using meteorological data and time series measurements of temperature improve understanding of variation in temperatures and mixing in lakes (MacIntyre and Melack 2009). Surface energy budgets indicate the amount of heat lost or gained via latent heat exchange (LE, evaporation), sensible heat exchange (SE, conduction), and net longwave (netLW = LW_{in} − LW_{out}) and shortwave (SW, corrected for reflection from the surface) radiation. The heat fluxes at a lake's surface can be expressed as Q_{tot} = netLW + netSW + LE + SE, where Q_{tot} is the surface energy flux.

An analysis of heat fluxes in Emerald Lake from 2014 through 2017, a range of years spanning winters with exceptionally wet and dry years, demonstrates the importance of snow as a mechanism driving differences in lake temperature between years, and highlights the relative importance of the other heat flux terms in dry years, when the influence of snow is reduced (Smits, MacIntyre, and Sadro 2020). During the ice-free period, net shortwave (netSW) was the largest term in the heat budget and accounted for the seasonal pattern of heating as well as the short-term variability associated with periods of cloud cover (figure 35). Net longwave radiation and latent heat exchange were the major heat loss fluxes. Sensible heat exchange was consistently the smallest component of the heat budget in both wet and dry years; however, the relative importance of other surface heat loss fluxes changed in relation to size of winter snowpack and duration of ice cover. The proportion of heat lost through net longwave radiation declined and the proportion lost through latent heat exchange increased in drier years, when the lake became ice-free earlier in the spring. Thus the size of winter snowpack is the principal factor regulating how warm Emerald Lake gets in the summer. In years with deep snowpacks, the lake remains colder primarily because ice cover blocks or reduces net shortwave radiation and prevents or reduces heating. In contrast, the effect of snowpack is reduced in dry years. Instead, spring and summer weather patterns drive heat fluxes and dictate lake warming patterns.

The meteorological and water temperature measurements necessary to compute heat budgets are rarely collected for Sierra lakes. However, summer measurements of temperatures, which are more common, reflect the role of both landscape position and lake morphometry in affecting water temperature. In lakes spanning a gradient in elevation, temperatures at 1 m were correlated with elevation, absorbance at 440 nm (Abs440), and the attenuation coefficient of photosynthetically available radiation (K_d PAR). Elevation affects heat budgets directly by altering downwelling solar radiation and indirectly by regulating the duration of ice cover and controlling vegetation and soil development. Vegetation cover increases with decreasing elevation and with it dissolved organic matter inputs to lakes (Sadro, Nelson, and

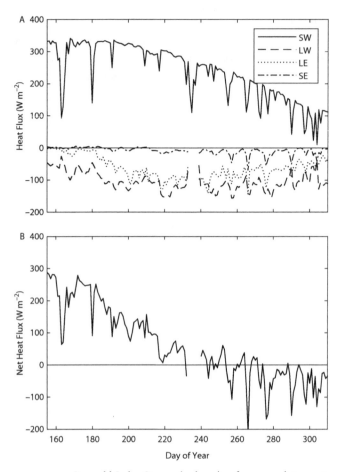

FIGURE 35. Emerald Lake: Energy budget (ice-free season) in 2016. Wind-sheltered Sierra lakes tend to gain energy through the summer and begin losing energy in the autumn when solar inputs decline and convective mixing increases the rate of cooling.

Melack 2012; Piovia-Scott et al. 2016). Consequently, increasing vegetation cover in catchments corresponds to increases in DOC, Abs440, and K_d PAR in lakes (figure 36). Because dissolved organic matter absorbs more strongly at short wavelengths, small changes in dissolved organic matter cause increased light attenuation and higher rates of warming within the upper water column. The increases in Abs440 and attenuation coefficient for PAR with increasing vegetation cover and decreasing elevation observed in the Sierra reflect these processes. A multiple regression model that included all three factors ($p < 0.05$ for all factors) described 95% of the

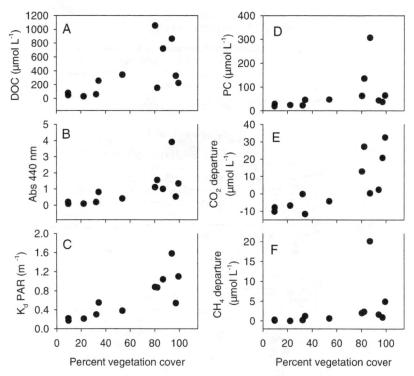

FIGURE 36. Elevational gradients in vegetation cover within watersheds are an important factor controlling the concentration of dissolved (A) and particulate organic matter (D) in lakes, the amount of colored dissolved organic matter (B) and attenuation coefficient of photosynthetically active radiation (C), and departure from saturation concentration for CO_2 (E) and CH_4 (F).

variation in August-September water temperature among a set of twelve lakes.

High-frequency measurements of water temperature made at the outlet of Emerald Lake closely tracked volume-weighted mean (VWM) lake temperatures over multiple years, indicating that measurements at the outlet provide a reliable index of lake temperatures. Ordinary least squares regression models that use outlet temperature to predict VWM lake temperature are fairly accurate during both the heating and cooling phases of the ice-free season: VWM temperature during heating phase (June–August) = 2.57 + outlet temperature (0.882) ($R^2 = 0.87$, $F_{114,1} = 778.18$, $p < 0.0001$); and VWM temperature during cooling phase (September–October) = 1.67 + outlet temperature (0.914) ($R^2 = 0.86$, $F_{123,1} = 746.78$, $p < 0.0001$). Volume-weighted mean lake temperatures in Emerald Lake range from 3° to 4° during the winter and up to ~20° in July

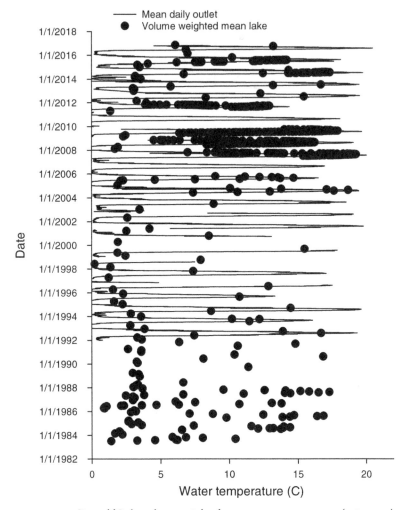

FIGURE 37. Emerald Lake volume-weighted mean water temperatures (1982–2012) and outlet water temperatures (1992–2012). Because of Sierra lakes' small size and comparatively high flushing rates, outlet temperatures in many of them provide a good measure of volume-weighted lake temperature for much of the year.

and August (figure 37). We examine long-term trends and variations in lake temperatures and causal factors in chapter 7.

Although heat loss through convective mixing typically dominates the Emerald Lake heat budget, winds can lead to upwelling and downwelling of the thermocline in stratified lakes, and Wedderburn (W) or Lake Number (L_N) indicates whether wind forcing at the surface is sufficient to tilt the thermocline.

$$W = g \, \Delta \, \rho \, h^2/(\rho \, u_w^{*2} \, L),$$

where:

g = gravity,

ρ = density,

$\Delta\rho$ = density difference across the thermocline,

h = depth of the mixed layer,

L = length of the lake, and

u_w^{*2} = water friction velocity derived from the shear stress.

W is the ratio of the buoyancy forces resisting mixing divided by the inertial forces that induce mixing times the aspect ratio of the lake. The Lake Number is an integral form of the Wedderburn Number. As explained in MacIntyre and Melack (2009), when these numbers are low (<< 1), a lake is mixed by wind; if near 1, the thermocline upwells to the surface; and if > 10, no tilting occurs. In Emerald Lake, wind speeds are usually less than 5 m s^{-1} and relative humidity averages 50% during the summer (Sadro, Melack, and MacIntyre 2011a). Lake Numbers during summer stratification in Emerald Lake exceed 100 and indicate the thermocline will not tilt and nonlinear internal waves will not form (MacIntyre and Melack 2009).

HYDROCHEMISTRY

In chapter 5 we examined the interactions among atmospheric deposition, hydrology, and watershed geochemistry that generate the solutes in the streams that enter Sierra lakes. Here we examine processes within the lakes that modify the chemical conditions and consider temporal and spatial variations in major solute chemistry in lakes based on time series measurements and Sierra-wide surveys. We begin with results from Emerald Lake, as representative of many high-elevation lakes in the Sierra Nevada.

Emerald Lake is chemically very dilute and weakly buffered, with a pH from 5.5 to 6.5 (figure 38), and calcium and bicarbonate are the two major ions. Mean concentrations (μEq L^{-1}) of anions and cations in the surface waters of Emerald Lake during August and September are as follows: chloride, 2.7; nitrate, 2.4; sulfate, 6.1; bicarbonate (assumed to be most of ANC), 28; calcium, 21; sodium, 11; magnesium, 3.9; and potassium, 2.9. Such values are expected in regions such as the Sierra Nevada, where weathering rates are

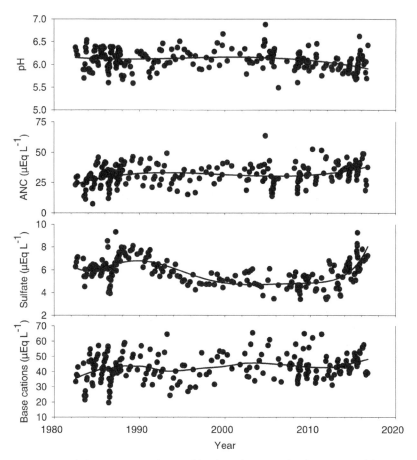

FIGURE 38. (A) Time series of Emerald Lake volume-weighted mean pH. (B) Time series of Emerald Lake volume-weighted mean acid neutralizing capacity (ANC). (C) Time series of Emerald Lake volume-weighted mean sulfate. (D) Time series of Emerald Lake volume-weighted mean sum of base cations (calcium, magnesium, sodium, potassium).

low and hydrology is dominated by snowmelt. Volume-weighted concentrations of acid neutralizing capacity (ANC) and sum of base cations (SBC) vary seasonally, with lowest concentrations during snowmelt and higher levels in autumn (figure 38). ANC varied from 2 µEq L^{-1} to 68 µEq L^{-1}, and SBC ranged from 20 µEq L^{-1} to 65 µEq L^{-1}. The lowest ANC occurred in July 1984 after an intense summer rainstorm (Engle and Melack 2001). Sulfate ranged from 4 µEq L^{-1} to 9 µEq L^{-1}, with higher values from the mid-1980s to the early 1990s and a slight increasing trend from 2011 to 2016; the earlier

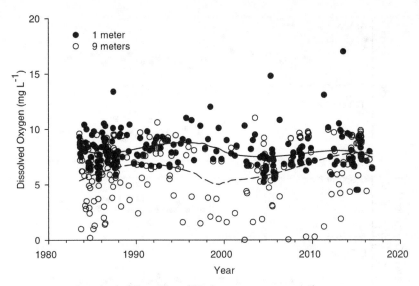

FIGURE 39. Dissolved oxygen. Emerald Lake, 1982–2016: 1 m and 9 m.

decline in lake sulfate paralleled reductions in atmospheric deposition of sulfate in the 1980s and 1990s (see figure 30). Further details about these patterns are provided by Melack, Hamilton, and Sickman (1993), Melack et al. (1993), and Melack et al. (1998).

Within-Lake Processes

Dissolved oxygen concentrations in lakes affect biogeochemical processes and when low, may cause stress or mortality among aquatic organisms. Measurements of dissolved oxygen within Emerald Lake do not indicate a long-term trend (figure 39). Concentrations in surface waters generally remained > 5 mg L^{-1} throughout the year; however, periodic hypoxia or anoxia developed in bottom waters during some winters. While episodes of complete anoxia in bottom waters were comparatively rare in Emerald Lake, or restricted to periods of winter ice cover, oxygen depletion was more frequent in bottom waters in some deeper lakes, where it occurred year-round (figure 40). Although the small number of lakes in our analysis precludes a broad assessment of dissolved oxygen dynamics, lake depth and basin morphometry appear to play roles in determining the frequency and duration of anoxia in Sierra lakes. Even in lakes in which anoxia at depth occurred comparatively often, surface waters remained well oxygenated year-round.

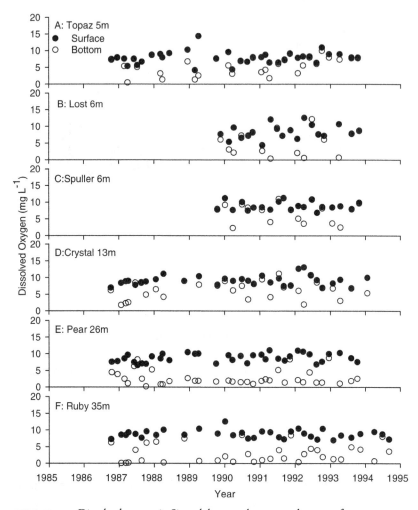

FIGURE 40. Dissolved oxygen in Sierra lakes near bottom and near surface.

Low dissolved oxygen concentrations, when they do occur, affect a number of important biogeochemical reactions. Reduced oxygen can enhance processes such as sulfate, nitrate, and iron reduction. These processes can contribute to alkalinity generation, though their significance will depend on concentrations of sulfate, nitrate, and iron and on residence time of water in the lake. Within-lake alkalinity production, in addition to ANC exported from the watershed, can contribute to acid neutralization in lakes and watersheds (Schindler et al. 1986; Psenner 1988).

Alkalinity in Gem Lake, a small high-elevation lake in the eastern Sierra Nevada, varied from as low as 30 μEq L^{-1} during spring snowmelt to values of > 240 μEq L^{-1} in autumn and winter as hydraulic residence times increased from ~7 d during snowmelt to about 135 d in winter (Stoddard 1987a). Using a mass balance approach to reconstruct weathering reactions in the basin, the increase in alkalinity was attributed to weathering in the watershed. In response to a suggestion that within-lake processes may also contribute to the alkalinity increase in Gem Lake (Schindler 1988; Kelly 1988), Stoddard (1988) calculated the production of alkalinity during the winter from sulfate reduction, nitrate retention, and cation exchange within the lake. Even with generous assumptions, these processes accounted for, at most, 1% of the total alkalinity flux.

To estimate the possible role of within-lake processes in influencing ionic composition of Emerald Lake, we selected periods with small watershed inputs and measured inputs and outputs to and from the lake: summer (late June to late October) and winter (December to mid-March), as described in detail by Melack, Cooper, and Jenkins (1989). During the summer, silica, base cations, nitrate, and sulfate in the outflow were lower than in the inflows, indicating uptake in the lake; no within-lake ANC flux was detected. No changes between inflows and the outflow were detected during winter.

Among potential factors influencing solute concentrations, groundwater seepage in Emerald Lake was determined to be very low (Kattelmann and Elder 1991) and to have no detectable influence. Solute fluxes between overlying water and fine sediments or submerged rocks were measured with benthic chambers (Melack, Cooper, and Holmes 1987; Melack, Cooper, and Jenkins 1989). The chambers were deployed over fine sediments by SCUBA to minimize disturbance. Shear velocities at the sediment-water interface were found to be similar inside the chambers, which had submersible pumps gently circulating the water, to velocities in the lake. All major solutes, except sodium and chloride, had detectable fluxes from fine sediments. Most of the ANC flux resulted from increases in ammonium, calcium, and potassium and decreases in sulfate and nitrate. The flux of silica from the sediments to the lake was similar to that for ANC. The submerged rocks (presumably with an epilithic biofilm) had detectable fluxes of ammonium and silica to the lake and uptake of nitrate. Our overall estimate of ANC flux from the sediments is 550 mEq m^{-2} y^{-1}. However, ammonium contributes to this flux and is likely to be oxidized in the water column or utilized by algae. If the ammonium flux is subtracted, the ANC flux is 270 mEq m^{-2} y^{-1}. Though the seasonal mass balances for the lake did not detect ANC fluxes, the direct measurements in benthic chambers indicate that the sediments are a source, though the magnitude may be overestimated by the chambers. For example, calculated fluxes of ANC, derived from base cations, based on gradients of interstitial pore water measured in Emerald Lake, are much less, though dependent on rough estimates of diffusion across the sediment-water interface (Amundson et al. 1988).

Based on field incubations in Emerald Lake to determine rates of phosphorus release from sediments, net consumption of phosphorus by sediments under oxic conditions occurred (Homyak, Sickman, and Melack 2014b). Under N_2-induced anoxia, zero or slightly positive net flux of phosphorus from the sediments occurred. Though phosphorus concentrations in sediments were sufficiently large to contribute to internal loading, sediments appear to be phosphorus sinks as a result of oxygenated waters overlying the sediments and of interactions with metal oxides. Emerald Lake's sediments contain a potentially active mass of 1,420 kg phosphorus, with 31% an exchangeable and reducible metal hydroxide pool (Homyak, Sickman, and Melack 2014b). However, high Al:P molar ratios found in most Sierra lakes are sufficient to prevent phosphorus desorption from anoxic sediments (Kopacek et al. 2005).

Solute fluxes associated with uptake by phytoplankton and benthic diatoms were estimated based on measurements of phytoplankton productivity and elemental stoichiometry of algal biomass. The decline of nitrate in the lake water parallels the calculated uptake by phytoplankton. Benthic diatom uptake of silica is likely to have led to the decrease in silica during the summer.

Temporal and Spatial Variations in Major Solute Chemistry

Measurements from eight Sierra lakes sampled after fall overturn over a span of fifteen years illustrate interannual variations in ANC. These lakes include four from the eastern side of the Sierra Nevada (Treasure and Gem Lakes in the Rock Creek basin, Upper Gaylor and Upper Granite Lakes near Tioga Pass) and four located in Sequoia National Park (Heather and Emerald Lakes in the Tokopah basin, Upper Mosquito and Upper Crystal Lakes in the Mineral King area). ANC in Emerald Lake ranged from 24 μEq L^{-1} to 38 μEq L^{-1}. The largest interannual variation was observed in Gem Lake (50 μEq L^{-1} to 200 μEq L^{-1}). Other lakes had a range of variation similar to that measured in Emerald Lake.

Starting in the early 1980s, as concern about the possible impacts of acidic atmospheric deposition in the Sierra Nevada arose (see chapters 4 and 5 and the section on acidification in this chapter), several surveys of solute chemical composition in lakes throughout the Sierra were done. The first survey with complete major ion analyses was done in 1981 and 1982 and comprised 73 lakes with several sampled up to four times (Melack et al. 1985). Engle and Melack (1997) summarize additional surveys done as part of studies funded by the California Air Resources Board (CARB). Other significant synoptic lake samplings include the 1999 repeat survey of 69 of the EPA's Western Lake Survey lakes (29 in the Sierra Nevada) by Clow et al. (2003); a survey of 95 lakes in the central Sierra Nevada in 2000 summarized by Berg et al. (2005); a rotating panel of 208 lakes in Sierra Nevada National Forest sampled during 2000–2009 presented by Shaw et al. (2013); a total of 82 lakes sampled from 14 catchments in 2006 by Sadro, Nelson, and Melack (2012); and 50 Sierra lakes sampled in a diatom calibration study during 2007 and 2008 by Sickman et al. (2013).

The only statistical sampling of lakes was performed by the U.S. EPA as part of their Western Lake Survey (WLS) (Landers et al. 1987) and was summarized and analyzed by Melack and Stoddard (1991). The WLS sampled 114 Sierra lakes of a possible population of 2,119 lakes. The population represented matches well the elevational distribution of lakes sampled by Melack, Stoddard, and Ochs (1985) and complementary CARB-funded studies. Median statistics for the population of WLS Sierra lakes are as follows: pH, 6.93; ANC, 60 μEq L^{-1}; SBC, 76 μEq L^{-1}; sulfate, 7 μEq L^{-1}; and nitrate, 0.4 μEq L^{-1}. These values are among the lowest obtained by the WLS, which included montane lakes in the Pacific Northwest; in the southern, central, and northern Rockies; as well as in California.

Watersheds in the Sierra Nevada often contain a series of hydrologically connected lakes arranged from high to lower elevation. Sadro, Nelson, and Melack (2012) sampled 14 different Sierra catchments containing a total of 82 lakes to examine the influences of a lake's position within the landscape on limnological characteristics, including major ion concentrations. They found the majority of variation in lake solute chemistry was associated with differences at the among-catchment spatial scale, most likely due to differences in mineral composition and soil development found at the landscape level. Variation in solute chemistry among lakes within the same catchment, in contrast, was lower, reflecting the common sources of weathering and hydrology.

Dissolved organic matter (DOM) is a fundamental aspect of lake ecosystems. It is often among the largest pools of carbon in lakes, and because colored dissolved organic matter absorbs solar radiation, especially in the ultraviolet range, it regulates the amount of light available through the water column and contributes to thermal dynamics. As a result, several ecosystem processes are affected by DOM, including rates of primary production and respiration, biogeochemical processes, stratification and mixing, dissolved oxygen concentrations, the feeding behavior of organisms, and food web structure. Both the concentration and the composition of the DOM pool affect these processes.

DOM in Sierra lakes is composed of a mixture of compounds of varying molecular weights and structures that come from two primary sources. It originates as organic matter fixed within a lake by aquatic primary producers (autochthonous) or as terrestrial organic matter that is subsequently flushed into lakes (allochthonous). Consequently, both lake trophic status and terrestrial loading determine DOM dynamics. The distinction between these two sources is important because autochthonous organic matter is often more labile and bioavailable for microbes, while terrestrial DOM comprises organic molecules that are more resistant to biodegradation.

The concentration and composition of DOM in Sierra lakes reflect variation in land cover that changes in relation to large-scale changes in elevation and small-scale heterogeneity within catchments (Sadro, Nelson, and Melack 2012). DOM concentrations are typically low, yet vary by over an order of magnitude. Dissolved organic carbon (DOC) ranged from 11 μmol L^{-1} to 317 μmol L^{-1}, dissolved organic phosphorus (DOP) from less than 0.1 μmol L^{-1} to 1 μmol L^{-1}, and dissolved organic nitrogen (DON) from 4 μmol L^{-1} to 18 μmol L^{-1}. Such variation is associated with elevation and gradients in vegetative cover and subsequent terrestrial inputs rather than lake productivity. For example, most high-elevation Sierra lakes are oligotrophic, with chlorophyll-a concentrations that average less than 1 μg L^{-1} and rarely climb above 3 μg L^{-1}. In contrast, the proportion of vegetative cover and the extent of soil development vary considerably with elevation, creating a large gradient in the pool of terrestrial organic matter available to enter lakes (see figure 36).

Elevation, slope, and the proportion of rock or vegetative cover in catchments are related to different aspects of organic matter in Sierra lakes. DOC and total fluorescence (TF) both increased relatively consistently along landscape gradients of increasing vegetative cover and decreasing elevation. Shrub cover accounted for 42% and 56% of the variability in DOC and TF, respectively, but only when excluding outlier catchments with abundant wet meadow or alpine tundra habitats (Sadro, Nelson, and Melack 2012). Such among-catchment variation reflects the high degree of landscape heterogeneity found at different nested scales within the Sierra Nevada. In contrast to the comparatively consistent within-catchment change in DOC and TF with vegetative cover, DON and DOP changed more through the alpine zone than through lower elevations.

Although a comprehensive biochemical assessment of DOM composition remains to be done for high-elevation Sierra lakes, a number of lines of evidence suggest composition changes in relation to variation in terrestrial sources. Excluding lakes in catchments with alpine tundra, the C:N ratio of DOM increases linearly with shrub cover (Sadro, Nelson, and Melack 2012), suggesting DOM in lakes at lower elevations becomes increasingly terrestrial in composition. The fluorescence index (FI), which is a relative measure of terrestrial and aquatic sources of fulvic acid in DOM (McKnight et al. 2001), varies predictably across lakes according to land cover classification types. Lakes with the largest terrestrial signal in DOM (FI ~1.25) were located in catchments where wetlands or alpine tundra were abundant, while lakes above tree line with little vegetation had the most autochthonous DOM (FI ~1.65). Falling between these end members and demonstrating an increasingly terrestrial signal in DOM composition with vegetative cover were lakes in subalpine catchments (FI ~1.55) and shrub-dominated habitats (FI ~1.45). Finally, the ratio between chlorophyll-a and TF, which reflects the balance between autochthonous and allochthonous source of DOM, decreases with increasing vegetative cover surrounding lakes. This suggests that DOM dynamics in higher-elevation lakes, where terrestrial inputs are low, are governed by phytoplankton to a larger extent than at lower elevations, where increasingly dense vegetation causes a shift toward terrestrial sources of organic matter.

Although many catchment characteristics change gradually with elevation, small-scale habitat heterogeneity can confound our ability to predict DOM at larger spatial scales. For example, models that predict the concentration or composition of DOM in lakes were more accurate within individual catchments, where up to 90% of variation was explained, than among catchment spatial scales (Sadro, Nelson, and Melack 2012). Much of the unexplained variation in DOM along elevation gradients can be attributed to landscape-level heterogeneity, such as the presence of specific habitat types, and patterns of hydrological connectivity among aquatic systems. In particular, wet meadow or tundra, which occurs across a broad elevation range, leaches high concentrations of terrestrial DOM into lakes.

ECOSYSTEM METABOLISM AND RATES OF PRIMARY PRODUCTION AND RESPIRATION

The creation and consumption of organic matter are fundamental ecosystem functions. Primary production and respiration are the foundation on which trophic structure is built. The rates of these processes have important implications for energy flow, biogeochemical cycling, and food web dynamics. At the ecosystem scale, the ratio between rates of gross primary production (GPP) and ecosystem respiration (ER) characterizes the balance between autotrophic and heterotrophic processes, providing an index of net ecosystem production (NEP).

Measurements of primary production in Emerald Lake were made between 1984–86 and 2007–9. Based on ^{14}C uptake rates in bottles of lake water collected from Emerald Lake, seasonal average rates of planktonic primary production made during the ice-free season in the mid-1980s were fairly low and ranged between 5 mmol C m^{-2} d^{-1} and 6 mmol C m^{-2} d^{-1}, with maximum seasonal averages up to 14 mmol C m^{-2} d^{-1} (Sickman and Melack 1992). More recent estimates of seasonal average GPP made using depth-integrated free-water measurements of dissolved oxygen ranged between 21 mmol C m^{-2} d^{-1} and 46 mmol C m^{-2} d^{-1} (Sadro, Melack, and MacIntyre 2011a, 2011b). Rate estimates that explicitly included spatial variability between littoral and pelagic habitats are slightly higher, about 55 mmol C m^{-2} d^{-1}, because they reflect the contribution of benthic production to whole-lake totals. The magnitude of these rates is consistent with those in oligotrophic lakes in California (Goldman, Jassby, and Powell 1989).

Respiration in Emerald Lake ranged between 15 mmol C m^{-2} d^{-1} and 48 mmol C m^{-2} d^{-1} during the ice-free seasons of 2007 to 2009 and is similar in magnitude to rates of primary production. Bacterioplankton accounted for 58% to 85% of total respiration over the growing season (Sadro, Nelson, and Melack 2011), reflecting a planktonic biomass dominated by bacterioplankton that is common in oligotrophic systems. The close agreement in estimates of GPP and ER (Sadro, Melack, and MacIntyre 2011a, 2011b) suggests that autotrophic and heterotrophic metabolisms are tightly coupled, underscoring bacterioplankton reliance on autotrophic production for much of their carbon.

Further evidence of tight autotrophic and heterotrophic coupling is provided by patterns of overnight oxygen drawdown in Emerald Lake in relation to diel changes in dissolved organic carbon concentrations (Sadro, Nelson, and Melack 2011). Perhaps reflecting an adaptation to a high light environment in a water column where attenuation of light is low, rates of respiration exceeded primary production in Emerald Lake as soon as the lake waters are no longer in direct sunlight. There was a logistic pattern in diel respiration rates, ER consistently higher during the evening and first few hours of night, followed by a rapid transition to lower background ER. In 2008, volumetric rates during the dusk phase averaged 1.35 μmol L^{-1} h^{-1}, an order of magnitude larger than the 0.11 μmol L^{-1} h^{-1} respiration rate that occurred through the remainder of the night. In addition to sustaining higher rates of respiration, DOC in water collected at dusk supported higher bacterioplankton community growth and remineralization rates than water collected at the end of night. Diel patterns in the drawdown of DOC (2.0 $\pm$ 0.8 μmol L^{-1}) that match oxygen-based estimates of the amount of DOC required to sustain the higher evening and early night rate phase of respiration (2.4 $\pm$ 1.3 μmol L^{-1}) suggest bacterioplankton are utilizing organic matter linked to daily primary production during the evening and early night before transitioning to less labile DOC during the remainder of the night. Together these data indicate that overnight variability in ER is driven by a bacterioplankton response to diel fluctuations in the quantity and possibly quality of DOC, most likely in response to the extracellular release of recently produced organic matter during the day.

The net ecosystem metabolism of Emerald Lake was slightly autotrophic over the ice-free season from 2007 to 2009, during which time NEP ranged between 4 mmol C m^{-2} d^{-1} and 10 mmol C m^{-2} d^{-1}. Positive NEP in an oligotrophic lake suggests autotrophy is possible in lakes where terrestrial inputs are low. Most high-elevation Sierra lakes are in catchments with thin soils that support comparatively low levels of terrestrial primary production. This constrains the magnitude of terrestrial loading to high Sierra lakes and contributes to their being autotrophic or in net metabolic balance during the ice-free season, despite having relatively low levels of primary production.

Although high-elevation lakes are likely to be autotrophic on average, there can be substantial spatial variability in metabolic rates within them. In Emerald Lake, primary production and respiration in shallow littoral habitats were up to four times higher than pelagic habitats (Sadro, Melack, and MacIntyre 2011b), reflecting the larger influence of benthic metabolism in shallow waters. The benthos in Emerald Lake was consistently heterotrophic. Despite high rates of benthic GPP, sediment ER was always higher, resulting in net benthic heterotrophy. The magnitude of benthic heterotrophy declined with depth, most likely as a result of declines in the abundance of benthic infauna. Within the water column, rates of GPP and ER tended to be elevated in association with a deep chlorophyll-a maxima located at or below the metalimnion, where localized particulate carbon and nitrogen concentrations were elevated (Sadro, Melack, and MacIntyre 2011a). Pelagic habitats contributed the majority of whole-lake metabolism (~75%), but benthic contributions reached as much as 50% in midsummer.

There are habitat-specific differences in factors regulating ecosystem metabolism in Emerald Lake. Metabolic rates in benthic incubations and free-water littoral areas were correlated with lake temperature; seasonal patterns of warming and cooling explained about 80% of this variability. In contrast, environmental control of epilimnetic metabolism was more complex. Rates of GPP were only weakly correlated with shortwave radiation, suggesting photosynthesis by phytoplankton may be light saturated. Rates of respiration were governed by the abundance of bacterioplankton in the water column and the quality of the organic matter available (C:N ratio of the seston). Overall net metabolic balance was governed by a combination of factors: seasonal shifts in DOC, bacterioplankton abundance, and changes in the quality of particulate organic matter. Although terrestrial inputs to Emerald Lake are comparatively low (Nelson, Sadrow, and Melack 2009), they are seasonally highest during the descending limb of the snowmelt hydrograph, when the fluorescence index for dissolved organic matter indicates predominantly allochthonous DOC and when phytoplankton biomass throughout the water column is seasonally lowest (Sadro, Melack, and MacIntyre 2011b). Under these conditions the epilimnion was predominantly heterotrophic. As snowmelt and terrestrial loading of DOC declined and phytoplankton biomass increased within the epilimnion, there was a gradual replacement of allochthonous with autochthonous DOC, changes that corresponded to a shift from the prevalence of early season heterotrophy to the fairly consistent autotrophy found during the remainder of the ice-free season.

A large autumn rainstorm in 2009 illustrates the sensitivity of high-elevation lakes to environmental variability (Sadro and Melack 2012). This atmospheric river–driven event simultaneously washed large amounts of terrestrial material into Emerald Lake and flushed phytoplankton from it, reducing chlorophyll-a concentrations by 40% and increasing DOC by 85%. Rates of whole-lake gross primary production dropped by nearly 50% after the flood, and ecosystem respiration increased by 30%. The net effect of the rainstorm was to cause a shift in net ecosystem production from typical autumn autotrophy (positive NEP) to heterotrophy (negative NEP). Infrequent events such as summer rains, winter floods, and avalanches striking Sierra lakes alter conditions in the lakes and their outlets (Kattelmann 1990; Williams and Clow 1990; Engle and Melack 2001).

Inferences from studies of metabolism in Emerald Lake suggest most high-elevation lakes should be near metabolic balance during the ice-free season because terrestrial organic matter inputs are low, and low nutrient concentrations limit aquatic primary production. However, the extent to which heterotrophy may increase with decreasing elevation remains unclear. Cohen and Melack (2020) extended the results from Emerald Lake with a comparative examination of metabolism and CO_2 fluxes to the atmosphere in five high-elevation lakes and five reservoirs on the eastern side of the Sierra Nevada. During summer, concentrations of dissolved CO_2 were above saturation and fluxes to the atmosphere were low. The length of ice cover was a predictor of summer surface CO_2. Net ecosystem production was variable between net autotrophy and net heterotrophy, across depth, site, and time. In lakes spanning a larger elevation gradient, the magnitude of heterotrophy is expected to increase with increasing terrestrial inputs. Data from twelve lakes spanning a large elevation gradient (A. P. Smits and S. Sadro, unpublished data), where particulate and dissolved organic carbon concentrations increased with vegetation cover (see figure 36), show summer CO_2 and CH_4 concentrations both increased substantially above saturation in lakes with increasing vegetation in their catchments and higher concentrations of DOC and POC.

AQUATIC ORGANISMS

Information about the species composition and abundance of aquatic organisms and their seasonal variations in Emerald Lake and other high-elevation Sierra lakes is limited to regional surveys and periods of a few years with detailed sampling, and long-term data sets are lacking, with the exception of introduced fish and amphibian surveys. As background for our next sections on eutrophication and acidification, we briefly summarize material available in a series of reports to the California Air Resources Board and relevant publications in scientific journals.

In Emerald Lake, phytoplankton peak in abundance in autumn and spring, with lowest numbers under ice (Melack, Cooper, and Holmes 1987). Algal groups include chlorophytes, dinophyceae, cryptophyceae, crysophytes, and cyanobacteria. Very small cells, with greatest axial linear dimension (GALD) < 5 μm, represented more than 50% of the cell counts. Nanoplankton (GALD 5 to 20 μm) were the second most abundant size class, followed by microplankton (GALD 20 to 64 μm) representing not more than 10%. Benthic diatoms were sampled from soft sediments and ceramic tiles in Emerald Lake. Sixty species were identified and evaluated in terms of their sensitivity to acidification; species lists and analyses are included in Melack, Cooper, and Holmes (1987). Quillwort (*Isoetes* cf. *bolanderi* var. *pygmaea*) grows on sandy substrata within the lake, and a moss (*Drepanocladus* sp.) occurs on mud midlake.

Nelson (2009) examined seasonal variation in the composition of bacterioplankton in Emerald Lake over three years. Bacterial clades were determined by coupling randomized clone sequence libraries to restriction fragment length polymorphism fingerprints. Bacterioplankton had recurring community types during spring snowmelt, ice-off, and fall overturn. Further work by Nelson et al. (2009) examined bacterioplankton composition using polymerase chain reaction-based phylogenetic fingerprinting in Emerald Lake and seventeen lake chains throughout the Sierra. Over three years, Emerald Lake's bacterioplankton community remained distinct from its inlet even during peak snowmelt. Among the lake chains, headwater lakes were less similar to their inlet streams than were downstream lakes, and bacterioplankton composition varied within catchments.

The zooplankton assemblage in Emerald Lake is dominated by a copepod (*Diaptomus signicauda*), three cladocerans (*Daphnia rosea, Holopedium gibberum, Bosmina longirostris*), and five rotifers (*Keratella taurocephala, K. quadrata, Polyarthra vulgaris, Conochilus unicornus, Trichocerca capucina*). Nearly all of the zooplankton species are most abundant during the summer (Melack, Cooper, and Holmes 1987). Samples of soft and hard substrata for benthic invertebrates contained an abundance of chironomids, water mites and sphaerid clams (*Pisidium* sp.); oligochaetes and chydorid cladocera were common, and low densities of a sponge (*Spongila lacustris*) occurred on and under hard substrata (Melack, Cooper, and Holmes 1987).

Context for the zooplankton found in Emerald Lake is provided by sampling of eight lakes and one vernal pond in the Tokopah basin (Melack,

Cooper, and Holmes 1987), data for seven Sierra lakes sampled for three to six years (Engle and Melack 1995), and Stoddard's (1987b) survey of seventy-five high-elevation lakes, all of which indicate that Emerald Lake's zooplankton are representative of Sierra lakes containing fish. Among the lakes and ponds of the Tokopah basin, 22 species of zooplankton were collected: two calanoid copepods, two cyclopoid copepods, ten cladocerans, and eight rotifers. Nine to ten of these species were collected in Emerald Lake. The number of species tended to decline with elevation (Spearman's r_s = −0.87). A large copepod (*Diaptomus eiseni*) and cladaceran (*Daphnia middendorffiana*) occurred only in fishless waters (e.g., Topaz Lake), while *Daphnia rosea* was abundant in all lakes with fish.

Stoddard (1987b) identified one common assemblage consisting of *Daphnia rosea*, *Diaptomus signicauda*, *Bosmina longirostris*, and *Holopedium gibberum*. Another assemblage included many of these same species plus *Cyclops vemalis*, *Diaphanasoma brachyurum*, *Polyphemus pediculis*, and/or *Ceriodaphnia affinis*. These two groups were associated statistically with fish presence and elevation. Two other assemblages, distinguished by fish absence and lake depth, contained *Daphnia middendorffiana* with either *Diaptomus shoshone* or *Diaptomus eiseni*. A few lakes lacked microcrustaceans or contained only *Chydorus* and/or *Alona* spp.

Amphibians observed in the Tokopah basin include mountain red-legged frogs (*Rana aurora*) and Pacific tree frog (*Hyla regilla*) larvae in fishless Topaz Lake and several ponds; Mount Lyell salamanders (*Hydromantes platycephalus*) have been sighted rarely. The historically widespread mountain yellow-legged frog (*Rana muscosa*) was not found around Emerald Lake and has declined catastrophically throughout the Sierra Nevada as a consequence of fish predation (Knapp and Matthews 2000), pesticides from lowland agricultural activities (Davidson and Knapp 2007), and the infectious disease chytridiomycosis, caused by a fungal pathogen (*Batrachochytrium dendrobatidis*) (Rachowicz et al. 2006). One consequence of the loss or decline of amphibians is fewer mountain garter snakes (*Thamnophis elegans elegans*), a predator of the frogs (Matthews, Knapp, and Pope 2002), and reduced bird abundances around lakes containing fish (Epanchin, Knapp, and Lawler 2010). Efforts to reintroduce the yellow-legged frog in the Tokopah basin have been unsuccessful.

The only fish species present in Emerald Lake is the brook trout (*Salvelinus fontinalis*); it was last stocked in 1959 and is now maintained through natural reproduction. A series of studies of brook trout in Emerald Lake and nearby lakes and streams investigated population sizes, survival, reproduction, growth, movements, size and age structure, and diets (Melack, Cooper, and Holmes 1987; Cooper et al. 1988; Melack, Cooper, and Jenkins 1989). During the mid-1980s Emerald Lake contained about 1,000 brook trout older than one year, with little annual variation. Chironomid pupae and larvae, cladocerans, and terrestrial insects dominated the fish diets in summer and autumn. Young-of-the-year varied from 428 to 67, with the low number associated with a large snowpack and late thaw in 1986. Considerable mortality of sac fry and embryos was noted after a large avalanche struck the lake in February 1986 and caused a scouring discharge through the outflow. In 1987 mortality of embryos occurred when the outlet temporarily stopped flowing. The result of these interannual hydrological differences contributed to the differences in the percentage of embryos surviving from the end of the spawning season to egg hatching: 86% (1985–86), 13% (1986–87), and 39% (1987–88). Emerald Lake and nearby lakes contain dense, reproducing populations of slow-growing, long-lived individuals with small adult body sizes. Migration of fish from lakes in the lower Tokopah basin to water bodies in the upper basin is prevented by the steep gradient of the Marble Fork River above the confluence with Pear Lake's outflow.

Impacts of Introduced Fish and Recovery after Fish Removal

Almost all lakes and ponds in the Sierra Nevada lacked fish historically as a result of barriers to upstream movements (Knapp 1996). However, as early as the mid-1800s introductions of trout (golden trout, *Oncorhynchus mykiss aguabonita*; rainbow trout, *O. mykiss*; brook trout, *Salvelinus fontinalis*; and brown trout, *Salmo trutta*) had begun (Moyle, Yoshiyama, and Knapp 1996). By the mid-1900s the majority of lakes contained fish, and the stocking was being done by the California Department of Fish and Game (now the California Department of Fish and Wildlife) and other organizations with policies and practices that changed over the years (Knapp, Corn, and Schindler 2001; Pister 2001). The occurrence of naturally reproducing populations of introduced trout in Sierra lakes appears to be fairly widespread based on surveys of trout populations in lakes without ongoing stocking (Armstrong and Knapp 2004).

Impacts of these trout introductions on native vertebrate and invertebrate species can be substantial, with vulnerability being high for amphibians, reptiles, conspicuous benthic invertebrates, and zooplankton and low for inconspicuous benthic invertebrates (Knapp et al. 2001). Harper-Smith et al. (2005) developed food webs including common species to illustrate the multiple interactions and impacts of introduced trout and recovery trajectories after stocking ended and fish were no longer present. Melack and Schladow (2016, figure 32.3) provide an artistically enhanced rendition of one of these food webs that highlights the complex effects of the addition of trout to the lakes.

Knapp, Matthews, and Sarnelle (2001) evaluated the response to introduced trout by comparing changes in selected animal groups in lakes not stocked to those with stocked populations. The mountain yellow-legged frog, conspicuous benthic macroinvertebrates, and large crustacean zooplankton (> 1 mm) were much less abundant in stocked lakes. In contrast, inconspicuous benthic invertebrate taxa, small crustacean zooplankton (< 1 mm), and rotifers (< 0.2 mm) had abundances similar to or greater than fishless lakes. Furthermore, introduced trout reduce linkages between aquatic and terrestrial habitats (Matthews, Knapp, and Pope 2002; Epanchin, Knapp, and Lawler 2010; Piovia-Scott et al. 2016).

Introduced fish may alter nutrient regeneration rates by consuming benthic or terrestrial organisms. To examine this possibility, Schindler, Knapp, and Leavitt (2001) applied a bioenergetic model and estimated that trout stocked in Sierra lakes increased recycling of phosphorus. However, based on whole-lake fish removal experiments, Sarnelle and Knapp (2005) found little direct influence by the trout on phytoplankton biomass and phosphorus limitation. In contrast, phytoplankton biomass and P limitation decreased after *Daphnia* reestablished.

Ecological trajectories of recovery following removal of fish indicate it takes 11 to 20 years before macroinvertebrates and zooplankton communities show characteristics that resemble fishless lakes (Knapp, Matthews, and Sarnelle 2001). Zooplankton capable of parthenogenetic reproduction (e.g., *Daphnia middendorffiana*) and benthic macroinvertebrates with strong-flying adults recovered more rapidly than benthic invertebrates with weaker flight abilities, such as mayflies. Recovery of obligate dioecious zooplankton (e.g., *Hesperodiaptomus shoshone*) was especially problematic (Sarnelle and Knapp 2004; Knapp et al. 2005; Knapp and Sarnelle 2008). Fish removal also resulted in mountain yellow-legged frog increasing in number in the lakes and its dispersal to nearby water bodies (Knapp, Boiano, and Vredenburg 2007). Though fish removal has a positive effect on the frogs, their long-term viability remains challenged by chytridiomycosis.

Phosphorus and nitrogen limitation of phytoplankton growth in the freshwaters of North America is widespread and has long been recognized as an important ecological process (Elser, Marzolf, and Goldman 1990; Stoddard et al. 2016). Several studies in the western United States have demonstrated nitrogen limitation of phytoplankton occurring in Lake Tahoe, Castle Lake in the Klamath Mountains, and lakes in the Rocky Mountains of Colorado (Goldman, Jassby, and Hackley 1993; Elser et al. 1995; Morris and Lewis 1988). The shift of Lake Tahoe phytoplankton from co-limitation by N and P to persistent P limitation has been attributed to increased nitrogen loading from runoff and atmospheric deposition (Jassby et al. 1994). Similar trends have been observed in lakes throughout the northern hemisphere (Elser et al. 2009). Limitation by iron has also been demonstrated in Lake Tahoe (Chang, Kuwabara, and Pasilis 1992). Camarero and Catalan (2012) reported that increased atmospheric phosphorus deposition to Pyrenean lakes over recent decades was causing phytoplankton to switch from being phosphorus-limited to being nitrogen-limited despite concurrent increased nitrogen deposition.

Emerald Lake is oligotrophic (Sickman and Melack 1991). Typical mid-summer concentrations of chlorophyll-a are around 1 μg L^{-1}, and particulate carbon and nitrogen levels are in the low micromolar range. Chlorophyll-a concentrations typically peak between 3 m and 6 m below the surface. During ice-free seasons, ammonium and phosphate concentrations are usually below detection in near-surface waters. Ammonium can reach higher levels near the bottom during the winter and 1 μmol L^{-1} to 2 μmol L^{-1} throughout the water column following the breakdown of stratification. Elevated nitrate concentrations during snowmelt (figure 41) are caused by a pulse of nitrate produced predominantly in catchment soils from subnivean mineralization and nitrification of organic matter (Sickman, Leydecker et al. 2003; see chapter 5). Mean nitrate concentrations in Emerald Lake ranged from < 0.1 μmol L^{-1} to 14.3 μmol L^{-1} and averaged 1.2 $\pm$ 3.3 μmol L^{-1} during the period from 1982 to 2016 (see figure 41).

Five of seven of experiments conducted in Emerald Lake during 1983, 1984, and 1987 showed that chlorophyll-a levels increased with phosphorus additions alone (Melack, Cooper, and Jenkins 1989). Similarly, particulate carbon and nitrogen increased significantly when phosphorus was added in three of five microcosm experiments. Nitrogen was not found to be the sole

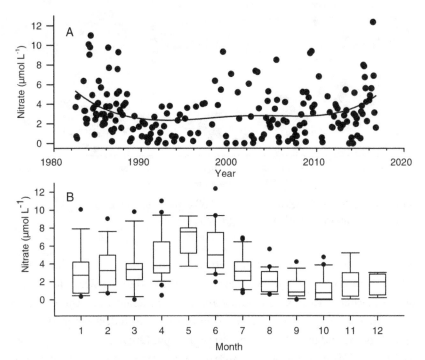

FIGURE 41. (A) Mean nitrate concentrations in Emerald Lake from 1982 to 2016. (B) Seasonal variations in nitrate concentrations in Emerald Lake during the period 1982–2016 displayed as box and whisker plots.

limiting nutrient and was co-limiting in only one experiment. The uptake of phosphate was large in comparison to the uptake of nitrate, as indicated by changes in nutrient concentrations in the enclosures. Results from mesocosm experiments conducted in 1985 (described below under Acidification) indicated that phytoplankton abundance was limited by phosphorus. Additions of phosphorus (in the presence of nitric acid) resulted in significantly greater concentrations of chlorophyll-a and particulate carbon and nitrogen, higher phytoplankton abundances, and increased uptake of ammonium compared to controls (Melack, Cooper, and Holmes 1987). Nitrate addition alone did not stimulate phytoplankton growth.

In contrast to the experiments conducted in the 1980s, time series measurements through 2016 indicate a tendency toward nitrogen limitation in the 1990s with periods of co-limitation and a return to periods of phosphorus limitation and others with co-limitation or nitrogen limitation after 2000 (figure 42). An analysis of trends in nutrient stoichiometry and particulate

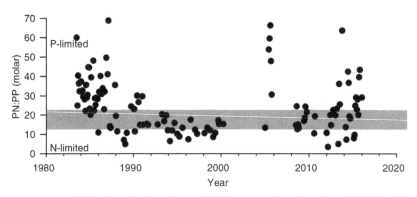

FIGURE 42. Ratio of particulate nitrogen (PN) to particulate phosphorus (PP). Shaded region represents co-limitation by nitrogen and phosphorus.

carbon concentrations within the seston found two lines of evidence that mild eutrophication of high-elevation lakes occurred in the Sierra during the 1990s (Sickman, Melack, and Clow 2003). In Emerald Lake there was an overall increase in particulate carbon concentrations, reflecting greater phytoplankton abundance, from 1991 through 1999. During that same period, PN:PP and DIN:TP ratios indicate the phytoplankton underwent a shift from phosphorus toward more frequent nitrogen limitation. These changes were reflected in results from synoptic surveys of lakes in 1985 and 1999 (~70% of lakes showed the trend) and were attributed to increased deposition and availability of phosphorus across the high-elevation Sierra. Recent evaluation of phosphorus dynamics within the Emerald Lake watershed suggest the majority of P-loading is from atmospheric deposition and subsequent processing through catchment soils (Homyak, Sickman, and Melack 2014a) rather than remineralization and release from lacustrine sediments (Homyak, Sickman, and Melack 2014b). Particulate carbon data from 2007 to 2013 reflect values similar to the mid-1980s, while dissolved and particulate nutrient ratios suggest a shift toward more frequent co- or phosphorus limitation of the phytoplankton.

In high-altitude Gem Lake, algal biomass was shown to be limited by phosphorus in combination with copper or iron (Stoddard 1987c). However, similar experiments conducted in Crystal, Emerald, Pear, Ruby, Spuller, and Topaz Lakes in 1992 found phytoplankton to be either nitrogen limited or co-limited by nitrogen and phosphorus (Sickman, Leydecker, and Melack 2001). Phytoplankton in Emerald Lake were co-limited by nitrogen and phosphorus in 2008 (Sadro 2011). There is likely to be a seasonal pattern to nutrient limitation within Sierra lakes resulting from the snowmelt-driven pulse in inorganic nitrogen, which while abundant in the spring declines to less than 1 μmol L^{-1} by late summer (see figure 41). In 2011, using bioassays we observed early season P-limitation (July 19–26), then a shift to co-limitation by N and P (August 2–9 and 24), and then N limitation by the end of summer (September 1).

Heard and Sickman (2016) conducted experimental bioassays in five Sierra Nevada lakes, including Emerald Lake, using a gradient of nitrogen concentrations and submersible fluorometer to frequently determine chlorophyll-a. Phytoplankton responded to nutrient additions (nitrogen and nitrogen plus phosphorus) in five of seven experiments, and the following effective doses (ED) were computed using dose-response curve-fitting: 10% ED range 0.33 μmol L^{-1} to 0.89 μmol L^{-1}, 50% ED range 1.0 μmol L^{-1} to 4.0 μmol L^{-1}, and the 90% ED range 3.1 μmol L^{-1} to 18 μmol L^{-1}. Applying these nutrient criteria to a set of 75 Sierra Nevada lakes sampled between 2008 and 2011 in Yosemite and Sequoia and Kings Canyon National Parks revealed that 20% to 30% of the lakes exceeded the 50% dose for nitrate and likely are responding to anthropogenic N deposition.

Bacteria also respond to nutrient addition or reduction and are important to the metabolism of organic matter in oligotrophic lakes. Nelson and Carlson (2011) evaluated responses of bacterioplankton to short-term enrichment by nitrate and phosphate in Emerald Lake. One experiment was conducted soon after ice-off and the second during autumn mixing. In both experiments, phosphate enrichment, but not nitrate addition, caused significant increases in bacterioplankton growth.

ACIDIFICATION

Lakes in the Sierra Nevada are sensitive to increased acidic deposition because their watersheds have a limited ability to neutralize acids owing to sparse vegetation, poorly developed soils, and slow-weathering granitic rock (chapter 5; see also Melack and Stoddard 1991; Melack et al. 1998). Steep slopes and high flushing rates during snowmelt limit the contact time between soils and snowmelt waters and increase the influence of atmospheric deposition on surface water chemistry. Huth et al. (2004) found that 10% to 15% of the outflow from the Emerald Lake watershed during snowmelt in 1998 passed directly from the snowpack into streams without interacting with soils and rocks.

Two mechanisms have been identified as causing depletion of ANC during snowmelt: acidification by strong acids and dilution by low ANC snowmelt (Williams, Brown, and Melack 1993). Acidification of surface waters by strong acids, primarily HNO_3 and H_2SO_4, from the melting snowpack was identified as a major cause of ANC depletion during the snowmelt seasons of 1986 and 1987 at Emerald Lake (Williams, Brown, and Melack 1993). In contrast, Leydecker, Sickman, and Melack (1999) concluded that episodic ANC depression occurs during snowmelt primarily as a result of dilution by naturally acidic snowmelt (pH ~5.6) rather than titration by strong acid anions.

One tool to analyze episodic acidification used by Leydecker, Sickman, and Melack (1999) compared ion concentrations during an episodic ANC minimum with concentrations before or after (Molot, Dillon, and LaZerte 1989). The analysis is derived from the ionic difference equation for ANC (all concentrations in $\mu Eq \cdot L^{-1}$):

$$ANC = SBC - SO_4^{2-} - NO_3^- - Cl^-$$

where SBC is the sum of base cations. Organic acids and monomeric aluminum were undetected in outflow and lake samples and are not included. The change in ANC concentration between a reference period and a minimum during an acidification episode is:

$$\Delta ANC = \Delta SBC + \Delta SO_4^{2-} + \Delta NO_3^- + \Delta Cl^-$$

Decreases in base cations (dilution) and increases in anion concentrations (acidification) both decrease ANC. If both sides of equation (2) are divided by ΔANC, and $\Delta(species)/(\Delta ANC)$ is defined as d(species), then the relative influences of dilution and acidification are expressed as:

$$dSBC + dSO_4^{2-} + dNO_3^- + dCl^- = 1$$

As explained by Leydecker, Sickman, and Melack (1999), concentrations from outflow and near-surface lake samples collected prior to the start of snowmelt provided an appropriate reference period, except for nitrate and sulfate. Nitrate concentrations, due to preferential elution from the snowpack and the flushing of nitrate from soils, peak early in snowmelt and decline to relatively low values by the ANC minimum. Thus Leydecker's approach minimizes the role of nitrate in early acidification. The role of sulfate in ANC depression is also underestimated because high sulfate concentrations are maintained throughout snowmelt as a result of soil desorption.

Another alternative approach, called superposition, is based on the Henriksen (1979) model of chronic acidification. Stoddard (1987) and Williams, Brown, and Melack (1993) substituted SBC for calcium and magnesium based on the assumption that the stoichiometric weathering of granitic materials yield HCO_3^- in equal proportion to the sum of base cations liberated. However, if the contributions of wet and dry deposition, the formation of cation-rich clays, the weathering of other minerals (e.g., pyrite), and catchment and lake biological processes are taken into account, the unacidified SBC/ANC ratio can vary considerably from unity. Leydecker, Sickman, and Melack (1999) applied this approach and reported that a third of the ANC depression in Emerald Lake was caused by acidification, while in Spuller Lake a large depression was caused almost solely by dilution. More generally, in shallow, short-residence-time lakes acidification accounts for < 10% of the ANC decrease (Lost, Spuller, and Topaz). In lakes where acidification caused 25% to 35% of the depression, the effects of dilution were reduced due to large lake volumes or slow snowmelt rates (Crystal, Emerald, Pear, and Ruby).

Experimental Studies of Biological Responses to Acidification

Most Sierra lakes contain taxa known to be sensitive to acidification (Holmes et al. 1989; Barmuta et al. 1990; Bradford et al. 1998). In particular, several invertebrate taxa are sensitive to acidic inputs. As found in our experiments, at pH below 5.6, the zooplankton *Daphnia rosea* and *Diaptomus signicauda* declined. Other zooplankton, *Bosmina longirostris* and *Keratella cochlearis*, initially increased in abundance but declined below pH of 5.

Experiments performed in large enclosures within lakes, often called mesocosms, permit replication in nearly natural conditions. To examine biological responses to acidification, a series of experiments were performed in mesocosms (volume, about 3,500–10,000 liters) suspended in Emerald Lake during the ice-free period (table 9; see also Melack, Cooper, and Holmes 1987; Barmuta et al. 1990).

Experiments 1 and 2 consisted of two levels of acid (nitric and sulfuric in equal proportions) and untreated controls; they tested the impact of acid alone on the plankton. A multifactorial design was used in Experiment 3, which consisted of 24 mesocosms assigned to six treatments or controls to differentiate the

TABLE 9. Design of mesocosm experiments. For each experimental treatment the number of replicate mesocosms is shown. Numbers in parentheses indicate the initial experimental pH of the acid treatments. The asterisks in Experiment 4 indicate the inclusion or exclusion of lake sediments in the control and acid addition enclosures in a cross-classified design (4 replicates per cell).

Treatment	Experiment 1	Experiment 2	Experiment 3	Experiment 4
Control	3 (6.3)	3 (6.3)	4 (6.3)	8* (6.3)
$H_2SO_4 + HNO_3$ (Hi-pH)	3 (4.4)	3 (5.2)	4 (5.2)	8* (5.6)
$H_2SO_4 + HNO_3$ (Lo-pH)	3 (3.8)	3 (4.1)		
HCl (Hi-pH)			2 (5.6)	
HCl (Lo-pH)			2 (5.0)	
$KNO_3 + Na_2SO_4$			4 (6.2)	
K_2HPO_4			4 (6.3)	
$K_2HPO_4 + H_2SO_4 + HNO_3$			4 (5.1)	

effects of ions associated with acidic deposition (hydrogen, nitrate, and sulfate) and nutrients (phosphate). In Experiment 4 one set of mesocosms extended into the sediments and one set included only the water column. Responses by phytoplankton, zooplankton, and zoobenthos were determined.

In Experiment 1, by day 15, chlorophyll-a was significantly ($p < 0.05$) greater in controls than in the Lo-pH treatment, but there were no significant particulate carbon (PC) or particulate nitrogen (PN) differences among treatments. In Experiment 2, owing to large within-treatment variability in phytoplankton biomass, there were no significant differences among treatments for PN, PC, or chlorophyll-a after 8 days. In Experiment 3, there were no a priori differences in PC, PN, or chlorophyll among treatments and no significant ($p > 0.05$) differences in PC concentration among treatments on any sampling date. By day 23, the N+S+H+P treatment had significantly higher ($p < 0.01$) concentrations of PN than the controls or the two HCl treatments, and chlorophyll-a concentrations were significantly higher in the N+S+H+P treatment ($p < 0.05$) than all other treatments and controls. Overall, the results from the mesocosm experiments indicate that moderate, short-term acidification alone had little effect on phytoplankton biomass. Severe acidification (pH < 4.0), did, however, cause temporary reductions in biomass.

Experiment 4 was designed to assess the role of sediments in response to episodic acidification. There were no initial differences in major ion chemistry, trace metals, or PC and PN among the treatments (table 10). By day 24 of the experiment pH had increased in both acid treatments; however, the increase was statistically greater in treatments including sediments ($p < 0.05$). PC and PN were not significantly different among treatments on day 0 or day 24 of the experiment. Ammonium concentrations were significantly higher ($p < 0.05$) in mesocosms with sediment compared to those without. ANC increased significantly ($p < 0.05$) in the acid treatments with sediments (indicating that ANC generation from the sediments titrated acid), whereas acid-treated mesocosms without sediments showed no change in ANC. The concentrations of aluminum and iron were not significantly different among treatments and were lower than those considered to be toxic to aquatic organisms (Havas 1985; Havas and Likens 1985).

Paleolimnological Evidence of Acidification

Analyses of microfossils in sediments deposited in Sierra lakes can extend information about the lakes in time. These analyses require establishing

TABLE 10. Pretreatment, initial, and final (days 0 and 24) chemical conditions in Experiment 4. All units are µEq L^{-1} except for aluminum and iron, which are µM. Mean values are shown for each treatment, with standard errors in parentheses. The two samples collected at different depths from each enclosure were averaged, and these averages were used to examine treatment effects on response variables. C_B is the sum of base cations (calcium, magnesium, sodium, potassium). NM = not measured.

Day 0

Treatment	H$^+$	NH$_4^+$	C$_B$	NO$_3^-$	SO$_4^{2-}$	ANC	Al	Fe
Acid + Sed.	0.55	1.3	38	1.5	6.1	34	0.6	0.4
	(0.03)	(0.3)	(0.4)	(0.03)	(0.04)	(1.3)	(0.08)	(0.08)
Acid + No Sed.	0.53	0.9	38	0.9	6.2	33	0.5	0.5
	(0.02)	(0.1)	(0.3)	(0.1)	(0.07)	(0.7)	(0.15)	(0.09)
No Acid + Sed.	0.48	0.8	37	1.4	6.1	33	0.8	0.6
	(0.02)	(0.05)	(0.7)	(0.03)	(0.05)	(0.7)	(0.15)	(0.10)
No Acid + No Sed.	0.55	0.8	38	1.4	6.1	35	0.7	0.5
	(0.02)	(0.07)	(0.5)	(0.04)	(0.03)	(0.9)	(0.18)	(0.10)

Day 23

Treatment	H$^+$	NH$_4^+$	C$_B$	NO$_3^-$	SO$_4^{2-}$	ANC	Al	Fe
Acid + Sed.	1.7	0.7	42	12	16	14	0.5	0.2
	(0.2)	(0.1)	(1.7)	(1.0)	(1.0)	(2.4)	(0.09)	(0.02)
Acid + No Sed.	2.1	0.2	43	15	18	6.5	0.7	0.2
	(0.2)	(0.05)	(0.9)	(0.3)	(0.2)	(0.8)	(0.13)	(0.02)
No Acid + Sed.	0.49	0.9	39	1.5	6.0	32	0.5	0.2
	(0.03)	(0.1)	(1.5)	(0.1)	(0.03)	(1.3)	(0.09)	(0.04)
No Acid + No Sed.	0.44	0.1	41	1.3	5.3	33	0.5	0.2
	(0.02)	(0.04)	(1.0)	(0.02)	(0.6)	(0.6)	(0.07)	(0.02)

relationships between the microfossils and environmental factors of interest. Several investigators have done so and established statistically reliable correlations between diatoms and pH, ANC, temperature, and salinity (Holmes, Whiting, and Stoddard 1989; Whiting et al. 1989; Bloom et al. 2003; Sickman et al. 2013) and between chironomid remains and temperature (Porinchu et al. 2007) based on sampling lakes distributed throughout the Sierra Nevada. When applied to age-dated cores of lacustrine sediments, records of paleolimnological conditions can be inferred.

Results from a paleolimnological study of diatoms in Emerald Lake's sediments indicate the lake has not experienced significant or permanent changes in ANC or pH since 1825 (Holmes, Whiting, and Stoddard 1989). Further diatom-inferred pH reconstructions from four Sierra lakes, including Emerald Lake, detected no strong trends but evidence of slight recent acidification at one site, subtle recent ANC increase at Emerald Lake, and slight long-term increases in ANC at Eastern Brook Lake (Whiting et al. 1989). Based on diatoms in a core from Moat Lake, located in the eastern Sierra Nevada, Sickman et al. (2013) inferred that ANC varied from 85 μEq L^{-1} to 115 μEq L^{-1} from ca. AD 600 to the early twentieth century, declined from the 1920s to the 1970s, and recovered between 1970 and 2005. Heard et al. (2014) examined spheroidal carbonaceous particles (SCPs) (an indicator of anthropogenic atmospheric deposition) and biogenic silica and δ^{13}C (productivity proxies) in sediments of Moat, Pear, and Emerald Lakes and found that reconstructed ANC was negatively correlated with SCPs and sulfur dioxide emissions. The decline in ANC from 1920 to 1970 was attributed to increasing acid deposition dominated by H$_2$SO$_4$ and then recovery of ANC from 1970 to 2005 caused by reduction of sulfur dioxide emissions resulting from the Clean Air Act. Complicating this interpretation is that climate warming also occurred during the same period, as inferred from chironomid remains (Porinchu et al. 2007). One consequence of warming could be earlier snowmelt and increased summer and autumn ANC (Sickman et al. 2013). The strong coherence, however, between declining sulfate concentrations in Emerald Lake and declining atmospheric deposition of sulfur during the 1980s and 1990s (see figure 30) shows that changes in atmospheric chemistry rather than climate change were the main drivers of recovery from acid deposition in the late twentieth century.

SEVEN

Variability, Trends, and Future Scenarios

Abstract. Long-term environmental change in the Sierra Nevada is linked to variability and trends in snowfall and solute deposition. Changes in snowfall and snowmelt caused by altered precipitation and energy fluxes modify the timing and quantity of hydrological and chemical fluxes, temperatures, and ecology of Sierra watersheds and lakes. Despite high rates of warming of annual air temperature in the Emerald watershed of $0.85°$ decade^{-1} over the past three decades, there is no long-term warming trend in lake temperature. This is because high interannual variability in snow water equivalent drives most of the variation through time in duration of snow cover, snowmelt dynamics, and lake temperature. Differences in the duration of ice cover and runoff between wet and dry years have important implications for biogeochemical and ecological responses, and locations where some precipitation falls as snow and some falls as rain will experience greater interannual variability in snow as the climate warms. During years with deep snowpacks, lakes have higher peak nitrate concentrations and lower acid neutralizing capacity. Dry years may increase sensitivity to acidification because of reduced weathering and lower atmospheric loading of cations but increase primary productivity in lakes by increasing residence times and extending the ice-free period.

Key Words. snowfall, snowmelt, energy fluxes, chemical fluxes, air temperature, water temperature, snow water equivalent, runoff, acid neutralizing capacity, nitrate

LARGE YEAR-TO-YEAR VARIABILITY IN CLIMATIC CONDITIONS is a hallmark of the Sierra Nevada, where "average" conditions are infrequently encountered. In this final chapter we examine integrated responses of lakes and watersheds to climate variability together with trends and future scenarios. Our measurements spanning three decades are combined with related historical data and climate projections. We show how changes in snowfall and snowmelt in response to altered precipitation and energy fluxes modify the timing and quantity of hydrological and chemical fluxes, temperatures, and ecology of Sierra watersheds and lakes.

By the 1990s, the effects of climatic variability and changes on inland waters throughout western North America were becoming evident (Williams, Losleben et al. 1996; Hauer et al. 1997; Melack, Dozier et al. 1997). For example, since about 1950, the proportion of total annual stream discharge occurring from April to July had decreased, while the proportion during autumn and winter had increased (Aguado et al. 1992), and winter and early spring runoff had increased during the period 1965–90 compared to 1939–64 (Pupacko 1993). On the eastern slope of the Sierra Nevada, five of the largest snowmelt floods since the 1920s occurred from 1978 to 1986, four of the lowest amounts of snowmelt occurred from 1987 to 1990 (Kattelmann 1992), and the most severe drought on record occurred from 2013 to 2015 (Margulis et al. 2016b), suggesting that extreme events could be increasing. Recent results by Berg and Hall (2015) support this suggestion. Floods are often associated with "atmospheric rivers," flows of warm water vapor that stretch eastward from the tropical Pacific Ocean. Climate predictions suggest these events are expected to increase in frequency in California (Dettinger 2011). In addition, statistical links between variability in precipitation and discharge and large-scale atmospheric patterns were becoming evident (Cayan and Peterson 1989; Redmond and Koch 1991; Dettinger and Cayan 1995; Cayan 1996;). A model of future precipitation (Marshall et al. 2019) indicates that in the elevations near the rain-snow transition, climate warming will cause an increase in the interquartile range of snowfall.

Long-term climate records are rare in the high-elevation Sierra Nevada. We first focus on Emerald Lake and nearby sites because our three decades of data allow detailed analyses. However, we recognize that climatic variability results from a combination of local and regional processes. Given the small-scale heterogeneity found in mountain ecosystems, our records must be used cautiously in the context of the whole Sierra Nevada and in relation to regional climate simulations, which do not match the scale at which our measurements were made.

CLIMATE WARMING TRENDS

We evaluated the magnitude of climate-related warming by comparing measurements of air temperatures made at three sites spanning a 1,200 m elevation

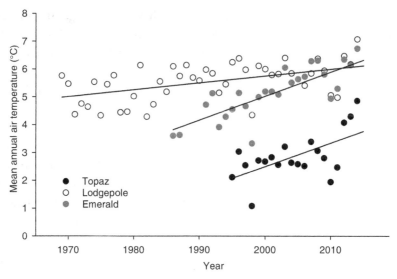

FIGURE 43. Mean annual air temperatures from three sites in the Tokopah basin: Lodgepole, Emerald, and Topaz.

gradient in the upper Marble Fork of the Kaweah River: Emerald and Topaz Lakes and Lodgepole (figure 43). Mean annual air temperature (± standard deviation) during the period when data are available for all sites (1986–2015) was 5.6 ± 0.9° at Lodgepole, 5.2 ± 0.9° at Emerald, and 2.9 ± 0.8° at Topaz, reflecting a pattern of decreasing temperature with increasing elevation. All three sites had significant average annual warming trends with rates of warming at higher elevations of 0.85° decade⁻¹, three times faster than the 0.25° decade⁻¹ rate of warming observed at the lower-elevation Lodgepole site.

Further examination of air temperature trends in the Emerald basin illustrates seasonal differences and variation between daytime and nighttime rates of warming. Warming trends were significant in all seasons except spring (figure 44), when the warming trend is somewhat masked by higher interannual variability. At 1.4° decade⁻¹, the rate of warming was highest during the summer. Autumn and winter rates of warming were 35% and 54% lower than the summer rate, respectively. Warming was evident for both daytime and nighttime mean temperatures, with nighttime rates of warming higher than daytime rates by 25% on average (20% to 26% depending on the season).

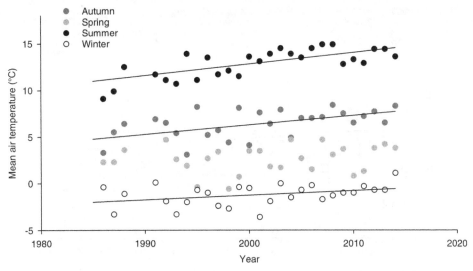

FIGURE 44. Seasonal variation in the rate of warming in air temperature in the Emerald basin. Mean temperatures for each season are plotted independently for summer, autumn, spring, and winter.

CHANGES IN PRECIPITATION

The Tokopah basin is becoming drier, and the interannual variation in precipitation is high (see chapter 4). Snow water equivalent has varied by more than a factor of 7, ranging from 442 mm to 3,177 mm between 1983 and 2016 in the Emerald watershed. Rainfall varied by over a factor of 20, ranging from 20 mm to 430 mm. Even so, SWE has had a weak decreasing trend of 3.4 mm year^{-1} over the last century in the Tokopah basin (figure 45). The decline has been larger in recent decades. Since 1970 SWE has declined by 14.3 mm, driven in large part by the high number of drought years between 2007 and 2016.

The proportion of precipitation falling as rain is increasing in the upper Marble Fork of the Kaweah River, including the Tokopah basin. Rain measurements made within the Emerald catchment and at the lower-elevation Lodgepole site illustrate both the temporal changes and how such changes vary with elevation. In the longer Lodgepole record, rain makes up 20% of total precipitation in comparison to 16% at Emerald. At both sites the percentage of rain varies depending on total amount of snowfall, ranging from 16% to 23% between wet and dry years at Lodgepole and between 10% and

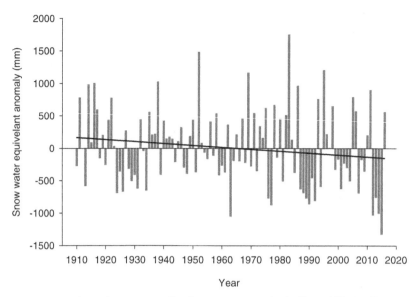

FIGURE 45. Annual snow anomalies from 1910 to 2015 in the Emerald basin. Bars are annual snow water equivalent (SWE) anomaly, based on snowpack density and depth measurements made near the beginning of April, typically the period of maximum accumulation. Measurements of SWE were made from 1985 to 2015 in the Emerald basin. SWE prior to 1985 is modeled from the Donner Summit record and other snow courses.

21% at Emerald. In the past decade, however, differences between the sites have declined despite the elevational variation. Prior to 2006, the difference between the sites in the amount of precipitation falling as rain was 5%, with more precipitation falling as rain at the lower-elevation Lodgepole site. However, in the past decade the sites have become more similar. Since 2006, the difference in the proportion of rain between the sites has dropped to about 1% and the frequency of years when the percent rainfall was higher at Emerald than at Lodgepole has increased from 20% to 50%.

These changes suggest that the proportion of precipitation falling as rain is increasing at these mountain sites and that the rate of change between the sites varies as a function of elevation. In the case of precipitation patterns, it is likely that the variability is related to changes in air temperature since the pattern is closely aligned in time with the changes in rates of warming seen between the two sites, and both the Lodgepole and Emerald sites have had an increase in the proportion of precipitation occurring when air temperatures are above 0 °C. Consequently, since 2006, the Emerald basin, which has been distinct from the Lodgepole site, has shifted to more closely resemble the lower-elevation site in terms of air temperature and precipitation patterns.

The global ambient atmospheric lapse rate is −6.5°C km⁻¹. If future temperature patterns approximately coincide with this value, then a change in storm temperatures of +1°C would raise the elevation of the 0°C isotherm by about 150 m. Thus, an increase in the elevation of the rain-snow transition is consistent with expectations in a warming climate. Harpold, Dettinger, and Rajagopal (2017) distinguish between "cold" and "warm" snow droughts. During "cold" snow droughts, snowfall is meager because of the lack of precipitation, whereas snowfall is sparse in "warm" snow droughts because much of the precipitation falls as rain instead. The drought in 2015 in the Sierra Nevada was characterized both by lack of precipitation and by warmer than normal temperatures; this combination led to an unprecedented loss of soil and groundwater, contributing to massive tree mortality at low to middle elevations in the Sierra (Fettig et al. 2019; Gouldin and Bales 2019; Argus et al. 2017), including the lower reaches of the Tokopah basin.

REGIONAL VARIATIONS AND TRENDS IN SWE

Concern over the effects of climate change has led to several studies examining half-century to century-long trends of the spring snowpack in the western United States and in the Sierra Nevada. Mote et al. (2005) concluded that the western mountains of the United States generally have a declining spring snowpack, especially since 1950. Their analysis with a coupled climate-hydrologic model showed that the decline is not sensitive to changes in land use or forest canopy. Changes were greatest in regions with mild winter temperatures such as the Cascade Mountains and the northern Sierra Nevada. Their analysis is less robust for the Sierra Nevada than for the rest of the western United States. Focusing specifically on the Sierra Nevada, Howat and Tulaczyk (2005) found a weak overall negative trend in spring SWE, accompanied by an increase in winter temperatures. The spatial distribution of their trend in SWE depends on both latitude and elevation, with increases in SWE at high elevations in the southern Sierra Nevada and decreases at lower elevations. Increased temperature and decreased precipitation are associated with SWE loss in the northern part of the range where peak elevations are lower.

An artifact of the use of operational snow data (snow courses and snow pillows) to examine trends is that the measurement sites do not cover the full range of elevations and the sampling sites are on flat terrain, as discussed in chapter 3. Hence the analyses of snow accumulation from the Tokopah basin are useful, even though the record is shorter (30 years) than for the snow courses (177 courses have more than 70 years of data) or the pillows (124 pillows have data records exceeding 30 years).

Similarly, analyses of remotely sensed data are necessarily of shorter duration than in situ records. The Landsat series of satellites dates to 1972, and the improved Thematic Mapper sensor was first included in 1982 on Landsat 4. The advantage of remote sensing, however, lies in the spatial and temporal coverage. The analysis by Margulis et al. (2016a, 2016b), described in chapter 3, showed, for example, that the 2015 drought in the Sierra Nevada was not only the driest on record, but the snow accumulation had a stronger elevational gradient than normal, so the drought was more severe at the lower elevations. In a future warmer climate, the distribution of the Sierra snow will be an important variable to monitor, in that lakes and other ecosystems will encounter different climates at the different elevations.

ICE COVER DURATION

Long-term trends in the duration of ice cover and the dates of ice formation and loss are documented for many lakes in North America and Europe (Magnuson et al. 2000). Such data are difficult to obtain for high-elevation lakes because people rarely reside near these lakes year-round. One of the few examples is observations over 33 years for seven alpine lakes in Colorado reported by Preston et al. (2016). They found that ice-off dates had shifted 7 days earlier and that snowfall in the spring was linked to the shift.

Using a time series of Landsat data from 1991 to 2013, Caillat and Melack (unpublished data) visually examined the images and estimated ice-off dates for 33 Sierra lakes ranging in altitude from ~2,200 m to ~3,900 m. Together, Landsats 5 and 7 acquired data every 8 days (assuming cloud-free conditions), so the temporal resolution of the dates is limited. For all lakes across all years, statistically significant ($p < 0.001$) relationships between the ice-out date and elevation and snow water equivalence were found. Although not statistically significant, a trend toward earlier ice-out was observed. In Emerald Lake, 81% of the variation in ice-out date ($p < 0.001$) was explained by interannual variation in snow water equivalent (Sadro et al. 2019).

Despite the high rates of warming air temperatures in the Tokopah basin, there was no warming trend in average annual water temperature in Emerald Lake over the past thirty years (figure 46A) (Sadro et al. 2019). Although lakes located in regions with warming air temperatures also tend to have warming trends in near-surface waters, numerous lakes behave differently (O'Reilly et al. 2015). The discrepancy in responses between air and lake temperatures at Emerald Lake suggests lake temperatures are being affected by factors other than those directly affecting air temperatures. Sadro et al. (2019) used principal component analysis (PCA) to determine relationships among climatic factors governing lake temperature and how those relationships vary seasonally. Annual snow water equivalent and seasonal mean values for lake temperature, short- and longwave radiation, air temperature, wind speed, and relative humidity were included in the analysis. The influence of large-scale processes such as the Pacific Decadal Oscillation or the El Nino/Southern Oscillation were examined but were not significantly correlated with lake temperature in any season.

The apparent decoupling in warming rates between Emerald Lake and its catchment is the result of snow and the mechanisms through which it affects water temperature that were discussed in chapter 6. Snowpack modulates lake temperature primarily by regulating the duration over which the lake absorbs solar radiation and secondarily through the influence of cold inflowing meltwaters. The effect of SWE across the 13° mean maximum summer temperature in Emerald Lake is largely linear (figure 46D). For every 100 mm decrease in SWE there was a 0.62° increase in lake temperature. Thus the high interannual variability in SWE largely drives the variation in lake temperature, and it is only in drought years, when the role of SWE is reduced, that the effect of other climatic factors dominates and a warming trend through time becomes evident (figure 46A, white symbols). For every degree increase in summer air temperature from climate warming during drought years, when the relative role of snow is small, there was a 0.65° increase in lake temperature (Sadro et al. 2019). The trend across drought years suggests that since 1990 water temperature in Emerald Lake has increased by 0.4° per decade as a result of climate warming, a pattern of increase consistent with global averages (O'Reilly et al. 2015).

Autumn differs from spring and summer in that lake temperatures are governed more by processes affecting heat loss than heat gain. Starting in

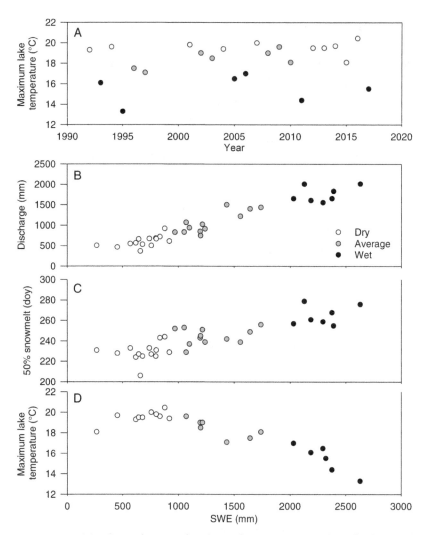

FIGURE 46. The accumulation and melting of winter snow regulates fundamental aspects of Sierra limnology. Lake temperatures have no warming trend across all years, but significant warming occurs during dry years when the effect of snow is low (A). SWE regulates both the magnitude (B) and the timing (C) of discharge. By controlling the duration of time that lakes are ice covered and can absorb solar radiation and the flushing rate during spring snowmelt, SWE drives the majority of variation in water temperature (D).

mid-September, climatic factors that affect water column mixing become important. Convective and wind-induced mixing are primary mechanisms of heat loss for lakes, as illustrated by the magnitude of the latent exchange in the lake energy budget (see figure 35). There were positive correlations between water temperature and the date on which the lake began mixing and shortwave radiation and negative correlations with wind speed and relative humidity (Sadro et al. 2019). During drought years SWE was comparatively unimportant in relation to the other climatic variables that together explained 97% of the variation in autumn lake temperature. The relative strength of regression coefficients suggests the effects of wind speed and air temperature were substantially larger than other factors, emphasizing their respective roles in autumn lake mixing.

Precipitation played the largest role in structuring lake temperature because SWE both directly impacted the lake heat budget and indirectly determined the influence of other climate variables on warming. Based on the loss of snow predicted by the RCP4.5 climate warming scenario, Sadro et al. (2019) estimated summer lake temperatures in the Sierra Nevada may warm 1.1°C to 10.5°C by the end of the twenty-first century, with the majority of lakes warming by at least 3.1°C. Larger temperature increases in lakes at low to mid-elevations are expected, as they will experience larger declines in SWE (Berg and Hall 2017; Fyfe et al. 2017). Other factors that vary across lakes but that were not considered in these estimates include wind exposure, local topographic relief, and light attenuation. The extent to which these factors may further mediate the effect of climate on water temperature remains unexplored.

SNOWMELT DYNAMICS

Many of the factors that govern snowmelt (see chapter 3) are expected to influence the snowmelt hydrograph (see chapter 4). In order to examine how snowmelt dynamics have changed in recent decades, Sadro et al. (2018) quantified a number of aspects of the snowmelt hydrograph related to the timing, magnitude, and rate of melt and evaluated them in relation to variation in specific climate factors. Their analysis confirms the overriding importance of the magnitude of snowfall on the snowmelt hydrograph (figure 46B, C), illustrates the extent to which variation in other meteorological factors affects patterns of snowmelt, and highlights the mechanisms through which long-term declines in mountain snowpacks will affect aquatic ecosystems.

SWE controlled most of the variation through time in snowmelt, and patterns were structured around differences between wet and dry years. As described in chapter 3, large snowpacks begin melting later and take longer to melt because of their mass, despite having faster melt rates than smaller snowpacks, which melt slower because they start melting earlier in the spring when energy inputs are lower and the length of the day is shorter. This apparent paradox of smaller snowpacks melting slower than larger snowpacks has been noted by Musselman et al. (2017). Data from snow pillows (see figure 9) support this contention. The date of snow disappearance depends both on the rate of melt and on the magnitude of the accumulation.

Hydrograph attributes related to the timing of melt (i.e., day on which melt begins and ends, reaches 50% cumulative discharge, and maximum discharge occurs) were positively correlated with total annual SWE and amount of snow falling in the spring, wind speed, and relative humidity and negatively correlated with short- and longwave radiation, air temperature, and percent of precipitation as rain. Spring snow maintains high albedo, delaying or extending the duration of melt. Spring rain contributes to the magnitude of runoff and occurs in conjunction with cloudy conditions that reduce energy inputs driving snowmelt. The cumulative effect of these differences meant that in dry years with smaller snowpacks, snowmelt began 25 days earlier and was 7 mm d^{-1} slower than in wet years with larger snowpacks. Moreover, because of the substantially higher discharge in years with large snowpacks, flushing rates in Emerald Lake were 2.5 times higher in wet years. Consequently, large differences in hydrological patterns between wet and dry years have important implications for biogeochemical and ecological responses.

HYDROCHEMICAL AND BIOLOGICAL RESPONSES TO VARIATIONS IN SWE AND SNOWMELT

High interannual variability in SWE has a strong influence on nitrate and acid neutralizing capacity in Sierra lakes. During years with deep snowpacks, lakes have higher peak nitrate concentrations and lower ANC (Sickman et al. 2003; Leydecker, Sickman, and Melack 1999). In particular, September ANC in Emerald Lake is negatively correlated with SWE (Sickman et al. 2013). These observations suggest that decreasing snowmelt volume or earlier snowmelt would result in higher ANC assuming no change in atmospheric deposition of acids (Sickman et al. 2003). Seasonal variation in the timing of peak nitrate concentrations was also affected by the magnitude of annual SWE, and nitrate increased during the rising limb of the hydrograph regardless of the amount of SWE (Sadro et al. 2018). In contrast, soluble reactive phosphorus concentrations were consistently low (< 0.1 μM), and seasonal patterns were lacking.

Sickman, Leydecker, and Melack (2001) and Sadro et al. (2018) considered the potential impacts of expected climate change on nitrate pulses and retention of nitrogen based on multiyear measurements at Emerald Lake. SWE was positively related to the timing of snowmelt, and the timing of snowmelt explained 90% of the variation in magnitude of the nitrate pulse, resulting in greater outflow nitrate concentrations when snowmelt occurred later in the year. Further, strong negative correlations occurred between nitrate retention and SWE and the timing of snowmelt (Sickman 2001; see also figure 32B). Years with earlier snowmelt, based on the date when 50% of annual runoff had occurred, have lower nitrate average concentrations and higher retention of dissolved inorganic nitrogen (DIN). Importantly, however, lake nitrate concentrations at the start of summer tend to be higher in dry years than in wet years, possibly because of lower flushing rates (Sadro et al. 2018). Hence if current trends toward warmer air temperatures and earlier snowmelt continue, nitrogen retention will increase in the Sierra Nevada. Higher nitrogen retention will result in lower nitrate pulses during snowmelt, but lower lake flushing rates may result in higher summer nitrate concentrations, partially as a result of dry deposition of nitrate. Long-term reductions in the Sierra snowpack are likely to reduce springtime episodic acid neutralizing capacity depression and lower overall sensitivity to nitrogen deposition (Leydecker, Sickman, and Melack 1999).

Increased snow depth and warmer winter temperatures forecast by some climate models could support higher rates of biogeochemical processes that recycle nitrogen and carbon in Sierra soils (Miller et al. 2007). The timing of the transition from anaerobic to aerobic processes (Sickman et al. 2003) could influence the nitrate pulse during snowmelt. Conversely, if later snowfall leads to frozen soils, anaerobiosis could occur and could determine whether soils at temperatures near melt are net producers or consumers of nitrogen.

In a multidecadal application of a linked hydrological and hydrochemical model, Meixner et al. (2004) found that chronic acidification of Emerald Lake is unlikely to occur at current rates of atmospheric deposition and that even episodic acidification would be quite limited with a doubling of atmospheric deposition. In contrast, scenarios of increases in acid deposition of

50%, 100%, and 150% were estimated by the empirical model of Leydecker, Sickman, and Melack (1999) to result in approximately 6%, 9%, and 14% of Sierra Nevada lakes (approximately 135, 185, and 290 lakes) becoming episodically acidified, that is, ANC < 0 μEq·L^{-1}, but no lakes becoming chronically acidified (ANC < 0 μEq·L^{-1}) at fall overturn. These differences may reflect that mineral weathering, and resulting ANC generation, is more responsive to changes in atmospheric deposition in the model used by Meixner et al. (2004). Furthermore, Meixner et al. (2004) found that variations in SWE from observed high to low values caused larger changes in ANC than doubling of atmospheric deposition alone. Our long-term atmospheric deposition measurements suggest that the concentration and loading of sulfuric acid has declined over the past thirty years, but the concentration of nitric acid may be increasing in the winter (see figure 22).

To assess the responses of chemical composition to an increased rate of snowmelt, as expected from global warming, Wolford and Bales (1996) applied a hydrological and geochemical model to the Emerald Lake watershed. Inputs during two contrasting years, one with high and one with low SWE, were used with conditions observed in 1986 and with snowmelt rates increased by 30%. During the period of snowmelt, modeled concentrations of calcium and ANC were lower with the higher snowmelt rate. Conversely, the higher rate of snowmelt caused the snow to disappear sooner, which led to low discharge in late summer and higher concentrations of solutes. When a 30% higher rate of snowmelt was combined with a doubled atmospheric deposition of solutes, ANC fell to near or below zero. Reductions of buffering capacity may also occur in extremely dry years. Solute balances from the Emerald Lake watershed study (see chapter 5) suggest that extreme drought conditions reduce weathering and atmospheric deposition of cations, resulting in net consumption of cations at the watershed scale and further reducing buffering capacity. In the dilute waters of the Sierra Nevada, these changes in solute concentrations may influence the aquatic biota, as was demonstrated experimentally in streams and lakes of the region (Barmuta et al. 1990; Kratz, Cooper, and Melack 1994).

Long-term trends in lake productivity and carbon cycling associated with changes in climate and altered hydrological variability are challenging to predict given the complex interactions involved among both biotic and abiotic drivers. However, studies from Emerald Lake suggest that changes in SWE will have important implications for lake productivity. Productivity responds to several factors linked to SWE, including the timing and magnitude of snowmelt, delivery of nutrients, changes in lake residence times, and shifts in the onset of lake warming (Sadro et al. 2018). Years with smaller snowpacks appear to create conditions more favorable for phytoplankton growth than years with large snowpacks. In years with low SWE, nitrate concentrations are higher at the start of summer at a time when water temperatures begin warming earlier and lake residence times are longer, resulting in higher particulate organic matter concentrations. Although the trends are not statistically strong, for every 1,000 mm decrease in SWE, particulate carbon (PC) increased by 6.4 μM and particulate nitrogen (PN) increased by 5.8 μM. Dissolved organic carbon (DOC) increased concomitantly with particulate organic matter (POM) in relation to declines in SWE, at a rate of 11.7 μM per increase per 1,000 mm decrease in SWE.

Several lines of evidence suggest particulate concentrations reflect lake productivity rather than terrestrial particulate matter inputs. First, if the mechanism were hydrological loading of terrestrial particulates, a positive trend would be expected rather than the negative trend observed. Second, the C:N ratio primarily ranges between 4 and 16 and does not have any trend with SWE, suggesting organic material is primarily of autochthonous origin. Finally, DOC and PC are correlated (r, 0.46) in Emerald Lake, and isotopic values suggest most DOC inputs during the summer stem from extracellular release from autochthonous primary producers (S. Sadro, unpublished data). Thus the mechanisms behind the observed trends in PC and PN with SWE are likely related to changes in primary productivity associated with variation in lake residence time and water temperature.

Our inferences that link concentrations of POM in Emerald Lake to actual rates of primary production are further supported by seasonal and spatial patterns in rates of gross primary production (GPP). As described in more detail in chapter 6, rates of pelagic and benthic GPP were influenced by water temperature and showed a strong pattern of seasonal increase with warming waters. Moreover, rates of GPP had a distinct spatial pattern throughout the water column, increasing with increasing concentrations of POM within the metalimnion. While the trends we observe in Emerald Lake suggest eutrophication will occur as a result of declines in snowpack, how variation in landscape characteristics or elevation will mediate loss of snow on lake productivity remains uncertain. Likewise, the extent to which predicted increases in storm intensity and intense runoff will interact with longer-term trends associated with declines in snowpack is unclear. As described in chapter 6, increased nutrient loading from moderate rain events tends to stimulate GPP in Emerald Lake, but large rain events can suppress GPP by flushing phytoplankton from the lake and increasing light attenuation through loading of colored DOM. Results from Emerald Lake add to evidence that changes in climate are affecting productivity in mountain lakes (Miller and McKnight 2015; Park et al. 2004; Preston et al. 2016); however, the implications for the trophic interactions, community structure, and food web energy flows of these changes all remain poorly understood for Sierra lakes. As discussed by Sadro et al. (2018), forecasting biological responses in mountain lakes will need to incorporate interactions and feedbacks among drivers.

FUTURE CLIMATE SCENARIOS

Projections of climate changes based on mathematical models that couple atmospheric, oceanic, and terrestrial components of Earth provide scenarios of possible future conditions. These models depend on the evolving understanding of the coupled Earth systems and how they respond to natural and anthropogenic perturbations. Though evaluated based on their fidelity to historical records, the real possibility of future conditions ranging outside the envelope of historical data complicates interpretations of future scenarios. While climate simulations generally project significant warming, projections of precipitation vary considerably, and predicting the probability of extremes, such as large SWE or prolonged droughts, is especially difficult. However, the expected transition of snow to rain at some elevations more likely results from warming temperatures, thus increasing the interannual variability in the elevations affected (Marshall et al. 2019) and the greater likelihood of warm snow droughts (Harpold, Dettinger, and Rajagopal 2017).

Advances in modeling future climate using nested models (Giorgi and Mearns 1991; Rauscher et al. 2008) and statistical downscaling (Pierce, Cayan, and Thrasher 2014) are especially relevant to the Sierra Nevada because of the importance of complex topography. Pierce and Cayan (2013) offer a useful example that provides further context for our examination of the Sierra Nevada. They compared the forecasts from thirteen global climate models downscaled to the western United States from 1950 to 2100 on a $1/8°$ × $1/8°$ latitude-longitude grid. Changes in response to these forecasts in several aspects of snowfall and accumulation, as determined by the variable infiltration capacity hydrological model, were examined. They selected the outputs from the Coupled Model Intercomparison Project, version 5 (CMIP5) archive and two greenhouse gas emission scenarios for the forecasts. Representative concentration pathways (RCP) numbers 4.5 and 8.5, approximating medium and high greenhouse gas emission scenarios, were used. They judged trends to be significant when the 95% confidence interval did not include zero.

Pierce and Cayan (2013) found that the models' forecasts varied considerably and differed with respect to different aspects of snow. For example, SWE (based on April 1 values) had the widest uncertainty range. In the Sierra Nevada they reported several results of interest: (1) not until 2060 or later did 80% of the models indicate a significant trend in snowfall; (2) the probability that SWE in the Sierra Nevada will decrease by at least 30% was 0.5 by the period 2040–69; and (3) in comparison to a historical period from 1976 to 2005, SWE at the end of the twenty-first century, based on the RCP 4.5 emissions scenario and the multimodel-ensemble average, decreased 30%. While representative of Sierra-wide conditions, there would be significant altitudinal and other geographic differences in SWE as the climate changes. A less optimistic future scenario based on RCP 8.5 (Marshall et al. 2019) shows a more dramatic change, with the frequency of SWE falling below the historical 25th percentile rising to over 40%.

As noted by Dettinger, Redmond, and Cayan (2004) and Lundquist et al. (2010), local processes influence the orographic precipitation that varies from storm to storm and season to season. Large-scale climatic data sets and simulations from climate change models do not capture these processes. Also, reductions in SWE can arise from earlier snowmelt, less snowfall, and the transition of precipitation from snow to rain.

CLIMATE CHANGE CHALLENGES AND FUTURE DIRECTIONS

As natural scientists who strive to determine mechanisms and make predictions and as citizens who share a deep spiritual connection with the mountains, the uncertainties of environmental changes remain a challenge. As the information gathered in this book demonstrates, there is much we know about the mechanistic linkages between climate and the ecosystem function of Sierra lakes. However, despite recent attempts to identify lakes most sensitive to changes in climate (Sadro et al. 2019), our ability to make specific predictions for many lakes remains hampered by uncertainty surrounding future changes in snowpack and a limited understanding of how climate forcing of lakes is mediated by small-scale variability in watershed characteristics, variation in lake morphometry, and community structure within lakes.

Thus, while much has been accomplished in recent decades, more remains to be done if we are to develop the ability to predict how Sierra lakes will change. Our research and the focus of this book has provided the basis for suggesting some responses of the Sierra lakes to environmental changes. We can estimate the likelihood of lake warming as a result of diminished ice cover. We know terrestrial inputs of organic matter will increase as the result of increased storm frequency and severity, altering the attenuation of light, affecting physical dynamics, and affecting ecosystem energetics. We know changes in the timing of snowmelt, nutrient concentrations, lake residence times, and the onset of warming will increase the productivity of some lakes. Less well understood is how the effect of snowpack on lake temperature is mediated by catchment or lake morphometric attributes that vary at landscape scales. Moreover, we lack an understanding of how differences in community structure within lakes interact with physical or chemical processes (most importantly, air quality and atmospheric deposition) to further mediate effects of climate change, which is vital for predicting population-level changes that may occur throughout the mountain range. Only by disentangling the relationships among these factors and others will we begin to develop a better ability to forecast responses in specific lakes. Our ability to make such predictions strengthens our basic scientific understanding of the mechanisms governing aquatic ecosystems and provides the tools necessary for management of Sierra Nevada lakes.

REFERENCES

Aber, J. D., W. McDowell, K. J. Nadelhoffer, A. Magill, G. Berntson, M. Kamakea, S. McNulty, W. Currie, L. Rustad, and I. Fernandez. 1998. Nitrogen saturation in temperate forest ecosystems: Hypothesis revisited. *Bioscience* 48: 921–34.

Aciego, S. M., C. S. Riebe, S. C. Hart, M. A. Blakowski, C. J. Carey, S. M. Aarons, N. C. Dove, J. K. Botthoff, K. W. W. Sims, and E. L. Aronson. 2017. Dust outpaces bedrock in nutrient supply to montane forest ecosystem. *Nature Comm.* 8, 14800. doi: 10.1038/ncomms14800.

Adams, A. 1979. *Yosemite and the Range of Light*. New York: Little, Brown.

Argus, D. F., F. W. Landerer, D. N. Wiese, H. R. Martens, Y. Fu, J. S. Famiglietti, B. F. Thomas, T. G. Farr, A. W. Moore, and M. M. Watkins. 2017. Sustained water loss in California's mountain ranges during severe drought from 2012 to 2015 inferred from GPS. *J. Geophys. Res.: Solid Earth* 122: 10559–85.

Aguado, E. 1990. Elevational and latitudinal patterns of snow accumulation departures from normal in the Sierra Nevada. *Theor. Appl. Clim.* 42: 177–85.

Aguado, E., D. Cayan, L. Riddle, and M. Roos. 1992. Climatic fluctuations and the timing of west coast streamflow. *J. Climate* 5: 1468–83.

Alagona, P. S., T. Paulson, A. B. Esch, and J. Marter-Kenyon. 2016. Population and land use. In H. Mooney and E. Zavaleta, eds., *Ecosystems of California: A Source Book*, 75–94. Oakland: University of California Press.

Amundson, R. G., J. Harte, H. Michaels, and E. Pendall. 1988. The role of sediments in controlling the chemistry of subalpine lakes in the Sierra Nevada, California. Final Report, Contract A4–042–32. California Air Resources Board.

Armstrong, P. 2014. *The Log of a Snow Survey: Skiing and Working in a Mountain Winter World*. n.p.: Abbott Press.

Armstrong, T. W., and R. A. Knapp. 2004. Response by trout populations in alpine lakes to an experimental halt to stocking. *Can. J. Fish. Aquat. Sci.* 61: 2025–37.

Bair, E. H., R. E. Davis, and J. Dozier. 2018. Hourly mass and snow energy balance measurements from Mammoth Mountain, CA, USA, 2011–2017. *Earth Syst. Sci. Data* 10: 549–63.

Bair, E. H., J. Dozier, R. E. Davis, M. T. Colee, and K. J. Claffey. 2015. CUES: A study site for measuring snowpack energy balance in the Sierra Nevada. *Frontiers Earth Sci.* 3: 58.

Bair, E. H., K. Rittger, R. E. Davis, T. H. Painter, and J. Dozier. 2016. Validating reconstruction of snow water equivalent in California's Sierra Nevada using measurements from the NASA Airborne Snow Observatory. *Water Resour. Res.* 52: 8437–60.

Bales, R. C., N. P. Molotch, T. H. Painter, M. D. Dettinger, R. Rice, and J. Dozier. 2006. Mountain hydrology of the western United States. *Water Resour. Res.* 42, W08432. doi: 10.1029/2005WR004387.

Barmuta, L. A., S. D. Cooper, S. K. Hamilton, K. W. Kratz, and J. M. Melack. 1990. Responses of zooplankton and zoobenthos to experimental acidification in a high-elevation lake (Sierra Nevada, California, U.S.A.). *Freshwater Biol.* 23: 571–86.

Barrett, A. 2003. National Operational Hydrologic Remote Sensing Center SNOw Data Assimiliation System (SNODAS) products at NSIDC. National Snow and Ice Data Center.

Bateman, P. C. 1961. Granitic formations in the east-central Sierra Nevada near Bishop, California. *Geol. Soc. Amer. Bull.* 72: 1521–38.

Bateman, P. C., B. W. Chappell, R. W. Kistler, D. L. Peck, and A. Busacca. 1988. Tuolumne Meadows Quadrangle-Analytic data. *U.S. Geological Survey Bull.* 1819. U.S. Geological Survey, Denver, CO.

Berg, N. H., A. Gallegos, T. Dell, J. Frazier, T. Proctor, J. O. Sickman, S. Grant, T. Blett, and M. Arbaugh. 2005. A screening procedure for identifying acid-sensitive lakes from catchment characteristics. *Environ. Monitor. Assess.* 105: 285–307.

Berg, N., and A. Hall. 2015. Increased interannual precipitation extremes over California under climate change. *J. Climate* 28: 6324–34.

Bieber, A. J., M. W. Williams, M. J. Johnsson, and T. C. Davinroy. 1998. Nitrogen transformations in alpine talus fields, Green Lakes Valley, Front Range, Colorado, USA. *Arctic Alpine Res.* 30: 266–71.

Blanchard, C. L., and K. A. Tonnessen. 1993. Precipitation-chemistry measurements from the California acid deposition monitoring program, 1985–1990. *Atmosph. Environ.* 27A: 1755–63.

Bloom, A. M., K. A. Moser, D. F. Porinchu, and G. M. MacDonald. 2003. Diatom-inference models for surface-water temperature and salinity developed from a 57-lake calibration set from the Sierra Nevada, California, USA. *J. Paleolimnol.* 29: 235–55.

Bohr, G. S., and E. Aguado. 2001. Use of April 1 SWE measurements as estimates of peak seasonal snowpack and total cold-season precipitation. *Water Resour. Res.* 37: 51–60.

Bradford, D. F., S. D. Cooper, T. M. Jenkins, K. Kratz, O. Sarnelle, and A. D. Brown. 1998. Influences of natural acidity and introduced fish on faunal assemblages in California alpine lakes. *Can. J. Fish. Aquat. Sci.* 55: 2478–91.

Brown, A. D., and L. J. Lund. 1991. Kinetics of chemical weathering in soils from a subalpine watershed. *Soil Sci. Soc. Amer. J.* 55: 1767–73.

————. 1994. Factors controlling throughfall characteristics at a high elevation Sierra Nevada site, California. *J. Environ. Qual.* 23: 844–50.

Brown, A. D., L. J. Lund, and M. A. Lueking. 1990. Integrated soil processes studies at Emerald Lake watershed. Final Report, Contract A5–204–32. California Air Resources Board.

Byron, E. R., R. P. Axler, and C. R. Goldman. 1991. Increased precipitation acidity in the central Sierra Nevada. *Atmosph. Environ.* 25A: 271–75.

Bytnerowicz, A., J. D. Burley, R. Cisneros, H. K. Preisler, S. Schilling, D. Schweizer, J. Ray, D. Dullen, C. Beck, and B. Auble. 2013. Surface ozone at the Devils Postpile National Monument receptor site during low and high wildland fire years. *Atmosph. Environ.* 65: 129–41.

Bytnerowicz, A., P. J. Dawson, C. L. Morrison, and M. P. Poe. 1991. Deposition of atmospheric ions to pine branches and surrogate surfaces in the vicinity of Emerald Lake, Sequoia National Park. *Atmosph. Environ.* 25A: 2203–10.

Bytnerowicz, A., and M. Fenn. 1996. Nitrogen deposition in California forests: A review. *Environ. Poll.* 92: 127–46.

Bytnerowicz A., M. Fenn, E. Allen, and R. Cisneros. 2016. Ecologically relevant atmospheric chemistry. In H. Mooney and E. Zavaleta, eds., *Ecosystems of California: A Source Book*, 107–28. Oakland: University of California Press.

Bytnerowicz, A., and D. M. Olszyk. 1988. Measurement of atmospheric dry deposition at Emerald Lake in Sequoia National Park. Final Report, Contract A7–32–039. California Air Resources Board.

Bytnerowicz, A., M. Tausz, R. Alonso, D. Jones, R. Johnson, and N. Grulke. 2002. Summer-time distribution of air pollutants in Sequoia National Park, California. *Environ. Poll.* 118: 187–203.

Camarero, L., and J. Catalan. 2012. Atmospheric phosphorus deposition may cause lakes to revert from phosphorus limitation back to nitrogen limitation. *Nature Comm.* doi:10.1038/ncomms2125.

Campbell, D. H., J. S. Baron, K. A. Tonnessen, P. D. Brooks, and P. F. Schuster. 2000. Controls on nitrogen flux in alpine/subalpine watersheds of Colorado. *Water Resour. Res.* 36: 37–47.

Campbell, D. H., C. Kendall, C. C. Y. Chang, S. R. Silva, and K. A. Tonnessen. 2002. Pathways for nitrate release from an alpine watershed: Determination using $\delta^{15}N$ and $\delta^{18}O$. *Water Resour. Res.* 38: 1–9.

Carroll, J. J., P. R. Miller, and J. Pronos. 2003. Historical perspectives on ambient ozone and its effects on the Sierra Nevada. In A. Bytnerowicz, M. J. Arbaugh, and R. Alonso, eds., *Ozone Air Pollution in the Sierra Nevada: Distribution and Effects on Forests*, 33–54. Developments in Environmental Science 2. Amsterdam: Elsevier.

Cayan, D. R. 1996. Interannual climate variability and snowpack in the western United States. *J. Climate* 9: 928–48.

Cayan, D. R., and D. H. Peterson. 1989. The influence of North Pacific atmospheric circulation on streamflow in the west. In D. H. Peterson, eds., *Aspects of Climatic Variability in the Pacific and the Western Americas*, 375–97. Geophysical Monograph Series 55. Washington, DC: American Geophyical Union.

Chang, C. C. Y., J. S. K. Kuwabara, and S. P. Pasilis. 1992. Phosphate and iron limitation of phytoplankton biomass in Lake Tahoe. *Can. J. Fish. Aquat. Sci.* 49: 1206–15.

Chow, J., J. G. Watson, D. H. Lowenthal, L.-W. A. Chen, and N. Motallebi. 2010. Black and organic carbon emissions: Review and applications for California. *J. Air Waste Manage. Assoc.* 60: 497–507.

———. 2011. PM2.5 source profiles for black and organic carbon emission inventories. *Atmosph. Environ.* 45: 5407–14.

Christenson, D. P. 1977. History of trout introductions in California high mountain lakes. In *Symposium on the Management of High Mountain Lakes in California's National Parks,* 9–15. Bethesda, MD: American Fisheries Society.

Church, J. E. 1933. Snow surveying: Its principles and possibilities. *Geog. Rev.* 23: 529–63.

Cline, D. W., R. C. Bales, and J. Dozier. 1998. Estimating the spatial distribution of snow in mountain basins using remote sensing and energy balance modeling. *Water Resour. Res.* 34: 1275–85.

Clow, D. 1987. Geologic controls on the neutralization of acid deposition and on the chemical evolution of surface and ground waters in the Emerald Lake watershed, Sequoia National Park, California. MSc thesis, California State University, Fresno.

Clow, D. W., M. A. Mast, and D. H. Campbell. 1996. Controls on surface water chemistry in the upper Merced River basin, Yosemite National Park, California. *Hydrol. Process.* 10: 727–46.

Clow, D. W., L. Nanus, K. L. Verdin, and J. Schmidt. 2012. Evaluation of SNODAS snow depth and snow water equivalent estimates for the Colorado Rocky Mountains, USA. *Hydrol. Process.* 26: 2583–91.

Clow, D. W., J. O. Sickman, R. G. Striegll, D. P. Krabbenhoft, J. G. Elliott, M. Dornblaser, D. A. Roth, and D. H. Campbell. 2003. Changes in the chemistry of lakes and precipitation in high-elevation National Parks in the Western United States, 1985–1999. *Water Resour. Res.* 39. doi:10.1029/2002WR001533.

Cohen, A., and J. M. Melack. 2020. Carbon dioxide supersaturation and fluxes to the atmosphere from high-elevation lakes and reservoirs in the Sierra Nevada, California. *Limnol. Oceanogr.* 65: 612–26.

Colbeck, S. C. 1981. A stimulation of the enrichment of atmospheric pollutants in snow cover runoff. *Water Resour. Res.* 17: 1383–88.

Cooper, S. D. 1989. An inventory of aquatic invertebrates found in streams of Sequoia National Park. In J. M. Melack, S. Hamilton, J. Sickman, and S. Cooper, *Effects of Atmospheric Deposition on Ecosystems in Sequoia National Park: Ecological Impacts on Aquatic Habitats,* chap. 3. Final Report, U.S. National Park Service, Cooperative Agreement No. 8006-2-0002.

Cooper, S. D., K. Kratz, R. W. Holmes, and J. M. Melack. 1988. An integrated watershed study: An investigation of the biota in the Emerald Lake system and stream channel experiments. Final Report, Contract A5-139-32. California Air Resources Board.

Cortés, A., S. MacIntyre, and S. Sadro. 2017. Flowpath and retention of snowmelt in an ice-covered arctic lake. *Limnol. Oceanogr.* 62: 2023–44.

Cosgrove, B. A., D. Lohmann, D. E. Mitchell, P. R. Houser, E. F. Wood, J. C. Schaake, A. Robock, C. Marshall, J. Sheffield, Q. Duan, L. Luo, R. W. Higgins, R. T. Pinker, J. R. Tarpley, and J. Meng. 2003. Real-time and retrospective forcing in the North American Land Data Assimilation System (NLDAS) project. *J. Geophys. Res.* 108: 8842.

Creamean, J. M., K. J. Suski, D. Rosenfeld, A. Cazorla, P. J. DeMott, R. C. Sullivan, A. B. White, F. M. Ralph, P. Minnis, J. M. Comstock, J. M. Tomlinson, and K. A. Prather. 2013. Dust and biological aerosols from the Sahara and Asia influence precipitation in the western U.S. *Science* 339: 1572–78.

Creed, I. F., and L. E. Band. 1998. Exploring functional similarity in the export of nitrate-N from forested catchments: A mechanistic modeling approach. *Water Resour. Res.* 34: 3079–93.

Crowley, B. E., P. L. Koch, and E. B. Davis. 2008. Stable isotope constraints on the elevation history of the Sierra Nevada Mountains, California. *Geol. Soc. Amer. Bull.* 120: 588–98.

Curtis, J. A., L. E. Flint, A. L. Flint, J. D. Lundquist, B. Hudgens, E. E. Boydston, and J. K. Young. 2014. Incorporating cold-air pooling into downscaled climate models increases potential refugia for snow-dependent species within the Sierra Nevada Ecoregion, CA. *PLoS ONE* 9: e106984.

Davidson, C., and R. A. Knapp. 2007. Multiple stressors and amphibian declines: Dual impacts of pesticides and fish on yellow-legged frogs. *Ecol. Appl.* 17: 587–97.

D'Azevedo, W. L. 1986. Washoe. In W. L. D'Azevedo, ed., *Handbook of North American Indians,* vol. 11, *Great Basin,* 466–98. Washington, DC: Smithsonian Institution.

Department of Water Resources. 2009. California Water Plan—Update 2009. Bulletin 160-09. Department of Water Resources, Sacramento, CA.

Dettinger, M. 2011. Climate change, atmospheric rivers, and floods in California: A multimodel analysis of storm frequency and magnitude changes. *J. Amer. Water Resour. Assoc.* 47: 514–23.

———. 2014. Climate change: Impacts in the third dimension. *Nature Geosci.* 7: 166–67.

Dettinger, M. D., and D. R. Cayan. 1995. Large-scale atmospheric forcing of recent trends toward early snowmelt runoff in California. *J. Climate* 8: 606–23.

Dettinger, M. D., F. M. Ralph, T. Das, P. J. Neiman, and D. R. Cayan. 2011. Atmospheric rivers, floods and the water resources of California. *Water* 3: 445–78.

Dettinger, M., K. Redmond, and D. Cayan. 2004. Winter orographic precipitation ratios in the Sierra Nevada: Large-scale atmospheric circulations and hydrologic consequences. *J. Hydrometeorology* 5: 1102–16.

Downing, J. A., Y. T. Praire, J. J. Cole, C. M. Duante, L. J. Tranvik, R. G. Stiegl, W. H. McDowell, P. Kortelainen, N. F. Caraco, J. M. Melack, and J. Middleburg. 2006. The global abundance and size distribution of lakes, ponds, and impoundments. *Limnol. Oceanogr.* 51: 2388–97.

Dozier, J. 1980. A clear-sky spectral solar radiation model for snow-covered mountainous terrain. *Water Resour. Res.* 16: 709–18.

———. 1989. Spectral signature of alpine snow cover from the Landsat Thematic Mapper. *Remote Sens. Environ.* 28: 9–22.

———. 2011. Mountain hydrology, snow color, and the fourth paradigm. *Eos* 92: 373–75.

Dozier, J., R. O. Green, A. W. Nolin, and T. H. Painter. 2009. Interpretation of snow properties from imaging spectrometry. *Remote Sens. Environ.* 113: S25–S37.

Dozier, J., and T. H. Painter. 2004. Multispectral and hyperspectral remote sensing of alpine snow properties. *Ann. Rev. Earth Planet. Sci.* 32: 465–94.

Dozier, J., T. H. Painter, K. Rittger, and J. E. Frew. 2008. Time-space continuity of daily maps of fractional snow cover and albedo from MODIS. *Adv. Water Resour.* 31: 1515–26.

Elder, K., J. Dozier, and J. Michaelsen. 1991. Snow accumulation and distribution in an alpine watershed. *Water Resour. Res.* 27: 1541–52.

Elser, J. J., T. Andersen, J. S. Baron, A.-K. Bergström, M. Jansson, M. Kyle, K. R. Nydick, L. Steger, and D. O. Hessen. 2009. Shifts in lake N:P stoichiometry and nutrient limitation driven by atmospheric nitrogen deposition. *Science* 236: 835–37.

Elser, J. J., F. S. Lubnow, E. R. Marzolf, M. T. Brett, G. Dion, and C. R. Goldman. 1995. Factors associated with interannual and intraannual variation in nutrient limitation of phytoplankton growth in Castle Lake, California. *Can. J. Fish. Aquat. Sci.* 52:93–104.

Elser, J. J., E. Marzolf, and C. R. Goldman. 1990. Phosphorus and nitrogen limitation of phytoplankton growth in the freshwaters of North America: A review and critique of experimental enrichments. *Can. J. Fish. Aquat. Sci.* 47: 1468–77.

Engle, D., and J. M. Melack. 1995. Zooplankton of high-elevation lakes of the Sierra Nevada, California: Potential effects of chronic and episodic acidification. *Arch. Hydrobiol.* 1331: 1–21.

———. 1997. Assessing the potential impact of acid deposition on high altitude aquatic ecosystems in California: Integrating ten years of investigation. Final Report, Contract 093–312. California Air Resources Board.

———. 2001. Ecological consequences of infrequent events in high-elevation lakes and streams of the Sierra Nevada, California. *Verh. Internat. Verein. Limnol.* 27: 3761–65.

Epanchin, P. N., R. A. Knapp, and S. P. Lawler. 2010. Nonnative trout impact an alpine-nesting bird by altering aquatic–insect subsidies. *Ecology* 91: 2406–15.

Eshleman, K. N., T. D. Davies, M. Tranter, and P. J. Wigington Jr. 1995. A two component mixing model for predicting regional episodic acidification of surface waters during spring snowmelt periods. *Water Resour. Res.* 31: 1011–21.

Farquhar, F. P. 1965. *History of the Sierra Nevada.* Berkeley: University of California Press.

Farr, T. G., P. A. Rosen, E. Caro, R. Crippen, R. Duren, S. Hensley, M. Kobrick, M. Paller, E. Rodriguez, L. Roth, D. Seal, S. Shaffer, J. Shimada, J. Umland,

M. Werner, M. Oskin, D. Burbank, and D. Alsdorf. 2007. The Shuttle Radar Topography Mission. *Rev. Geophys.* 45: RG2004.

Felzer, R. 1981. *High Sierra Hiking Guide No. 8, Mineral King.* Berkeley, CA: Wilderness Press.

Fenn, M. E., A. Bytnerowicz, and D. Liptzin. 2012. Nationwide maps of atmospheric deposition are highly skewed when based solely on wet deposition. *BioScience* 62: 621.

Fenn, M. E., R. Haebuer, G. S. Tonnesen, J. S. Baron, S. Grossman-Clarke, D. Hope, D. A. Jaffe, S. Copeland, L Geiser, H. M. Rueth, and J. O. Sickman. 2003. Nitrogen emissions, deposition and monitoring in the western United States. *BioScience* 53: 391–403.

Fettig, C. J., L. A. Mortenson, B. M. Bulaon, and P. B. Foulk. 2019. Tree mortality following drought in the central and southern Sierra Nevada, California, US. *Forest Ecol. Manage.* 432: 164–78.

Field, C. B., G. C. Daily, F. W. Davis, S. Gaines, P. A. Matson, J. Melack, and N. L. Miller. 1999. *Confronting Climate Change in California: Ecological Impacts on the Golden State.* Washington, DC: Union of Concerned Scientists and Ecological Society of America.

Finley, J. B., and J. I. Drever. 1997. Chemical mass balance and rates of mineral weathering in a high-elevation catchment, West Glacier Lake, Wyoming. *Hydrol. Process.* 11: 745–64.

Flerchinger, G. N., W. Xaio, D. Marks, T. J. Sauer, and Q. Yu. 2009. Comparison of algorithms for incoming atmospheric long-wave radiation. *Water Resour. Res.* 45: W03423.

Fowler, C. S., and S. Liljeblad. 1986. Northern Paiute. In W. L. D'Azevedo, ed., *Handbook of North American Indians,* vol. 11, *Great Basin,* 435–66. Washington, DC: Smithsonian Institution.

Frączek, W., A. Bytnerowicz, and M. J. Arbaugh. 2003. Use of geostatistics to estimate surface ozone patterns. In A. Bytnerowicz, M. Arbaugh, and R. Alonso, eds., *Ozone Air Pollution in the Sierra Nevada: Distribution and Effects on Forests,* 215–47. Developments in Environmental Science 2. Amsterdam: Elsevier.

Fyfe, J. C., C. Derksen, L. Mudryk, G. M. Flato, B. D. Santer, N. C. Swart, N. P. Molotch, X. Zhang, H. Wan, V. K. Arora, J. Scinocca, and Y. Jiao. 2017. Large near-term projected snowpack loss over the western United States. *Nature Comm.* 8: 14996. doi:10.1038/ncomms14996.

Garrels, R. M., and F. T. Mackenzie. 1967. Origin of the compositions of some springs and lakes. In W. Stumm, ed., *Equilibrium Concepts in Natural Water Systems,* 222–42. Advanced Chemistry Series 67. Washington, DC: American Chemical Society.

Giorgi, F., and L. O. Mearns. 1991. Approaches to the simulation of regional climate change: A review. *Rev. Geophys.* 29: 191–216.

Goldman, C. R., A. D. Jassby, and S. H. Hackley. 1993. Decadal, interannual, and seasonal variability in enrichment bioassays at Lake Tahoe, California-Nevada, USA. *Can. J. Fish. Aquat. Sci.* 50: 1489–96.

Goldman, C. R., A. Jassby, and T. Powell. 1989. Interannual fluctuations in primary production: Meteorological forcing at two subalpine lakes. *Limnol. Oceanogr.* 34: 310–23.

Goulden, M. L., and R. C. Bales. 2019. California forest die-off linked to multi-year deep soil drying in 2012–2015 drought. *Nature Geosci.* 12: 632–37.

Green, T. 2018. *Carleton Watkins: Making the West American.* Oakland: University of California Press.

Greene, E., K. Birkeland, K. Elder, I. McCammon, M. Staples, and D. Sharaf. 2016. *Snow, Weather, and Avalanches: Observational Guidelines for Avalanche Programs in the United States.* 3rd ed. Bozeman, MT: American Avalanche Association.

Gruell, G. E. 2001. *Fire in Sierra Nevada Forests: A Photographic Interpretation of Ecological Change since 1849.* Missoula, MT: Mountain Press.

Hadley, O. L., C. E. Corrigan, T. W. Kirchstetter, S. S. Cliff, and V. Ramanathan. 2010. Measured black carbon deposition on the Sierra Nevada snow pack and implications for snow pack retreat. *Atmosph. Chem. Phys.* 10: 7505–13.

Hall, D. K., G. A. Riggs, V. V. Salomonson, N. E. DiGirolamo, and K. J. Bayr. 2002. MODIS snow-cover products. *Remote Sens. Environ.* 83: 181–94.

Hardy, J. P., R. Melloh, G. Koenig, D. Marks, A. Winstral, J. W. Pomeroy, and T. Link. 2004. Solar radiation transmission through conifer canopies. *Agric. For. Meteorol.* 126: 257–70.

Harper-Smith, S., E. L. Berlow, R. A. Knapp, R. J. Williams, and N. D. Martinez. 2005. Communicating ecology through food webs: Visualizing and quantifying the effects of stocking alpine lakes with trout. In P. DeRuiter, J. C. Moore, and V. Wolters, eds., *Dynamic Webs: Multispecies Assemblages, Ecosystem Development, and Environmental Change,* 407–23. New York: Elsevier/Academic Press.

Harpold, A. A., M. Dettinger, and S. Rajagopal. 2017. Defining snow drought and why it matters. *Eos* 98.

Harrington, R. F., and R. C. Bales. 1998a. Interannual, seasonal, and spatial patterns of meltwater and solute fluxes in a seasonal snowpack. *Water Resour. Res.* 34: 823–831.

———. 1998b. Modeling ionic solute transport in melting snow. *Water Resour. Res.* 34: 1727–36.

Hauer, F. R., J. S. Baron, D. H. Campbell, K. D. Fausch, S. W. Hostetler, G. H. Leavesley, P. R. Leavitt, D. M. McKnight, and J. A. Stanford. 1997. Assessment of climate change and freshwater ecosystems of the Rocky Mountain, USA and Canada. *Hydrol. Process.* 11: 903–24.

Havas, M. 1985. Aluminum bioaccumulation and toxicity to *Daphnia magna* in softwater at low pH. *Can. J. Fish. Aquat. Sci.* 42: 1741–48.

Havas, M., and G. E. Likens. 1985. Toxicity of aluminum and hydrogen ions to *Daphnia catawba, Holopedium gibberum, Chaoborus punctipennis* and *Chironomus anthrocinus* from Mirror Lake, New Hampshire. *Can. J. Zool.* 63: 1114–19.

Heard, A. M. 2013. Global change and mountain lakes: Establishing nutrient criteria and critical loads for Sierra Nevada lakes. PhD dissertation, University of California, Riverside.

Heard, A. M., and J. O. Sickman. 2016. Nitrogen assessment points: Development and application to high-elevation lakes in the Sierra Nevada, California. *Ecosphere* 7: e01586.10.1002/ecs2.1586.

Heard, A. M., J. O. Sickman, N. L. Rose, D. M. Bennett, D. M. Lucero, J. M. Melack, and J. H. Curtis. 2014. Twentieth-century atmospheric deposition and acidification trends in lakes of the Sierra Nevada, California (USA). *Environ. Sci. Tech.* 48: 10054–61.

Hedrick, A., H.-P. Marshall, A. Winstral, K. Elder, S. Yueh, and D. Cline. 2015. Independent evaluation of the SNODAS snow depth product using regional-scale LiDAR-derived measurements. *Cryosphere* 9: 13–23.

Hedrick, A. R., D. Marks, S. Havens, M. Robertson, M. Johnson, M. Sandusky, H.-P. Marshall, P. R. Kormos, K. J. Bormann, and T. H. Painter. 2018. Direct insertion of NASA Airborne Snow Observatory–derived snow depth time series into the iSnobal energy balance snow model. *Water Resour. Res.* 54: 8045–63.

Henriksen, A. 1979. A simple approach for identifying and measuring acidification of freshwater. *Nature* 278: 5442–545.

Herbst, D. B., M. T. Bogan, S. K. Roll, and H. D. Safford. 2011. Effects of livestock exclusion on in-stream habitat and benthic invertebrate assemblages in montane streams. *Freshwater Biol.* 57: 204–17.

Herbst, D. B., E. L. Silldorff, and S. D. Cooper. 2009. The influence of introduced trout on the benthic communities of paired headwater streams in the Sierra Nevada of California. *Freshwater Biol.* 54: 1324–42.

Herschy, R. W. 1978. *Hydrometry.* New York: John Wiley and Sons.

Hicks, B. B., P. P. Hosker Jr., T. P. Meyers, and J. D. Womack. 1991. Dry deposition inferential measurement techniques I. Design and tests of a prototype meteorological and chemical system for determining dry deposition. *Atmosph. Environ.* 25: 2345–59.

Hill, M. 1975. *Geology of the Sierra Nevada.* Berkeley: University of California Press.

Hollstein, A., K. Segl, L. Guanter, M. Brell, and M. Enesco. 2016. Ready-to-use methods for the detection of clouds, cirrus, snow, shadow, water and clear sky pixels in Sentinel-2 MSI images. *Remote Sens.* 8: 666.

Holmes, R. W., M. C. Whiting, and J. L. Stoddard. 1989. Diatom-inferred pH and acid neutralizing capacity changes since 1825 A.D. in a dilute, high elevation, Sierra Nevada lake. *Freshwater Biol.* 21: 295–310.

Homan, J. W., C. H. Luce, J. P. McNamara, and N. F. Glenn. 2011. Improvement of distributed snowmelt energy balance modeling with MODIS-based NDSI-derived fractional snow-covered area data. *Hydrol. Process.* 25: 650–60.

Homyak, P. M., M. Kamiyama, J. O. Sickman, and J. P. Schimel. 2016. Acidity and organic matter promote abiotic nitric oxide production in drying soils. *Global Change Biol.* 23: 1735–47.

Homyak, P. M., J. O. Sickman, and J. M. Melack. 2014a. Pools, transformations and sources of P in high-elevation soils: Implications for nutrient transfer to Sierra Nevada lakes. *Geoderma* 217–18: 65–73.

———. 2014b. Sediment phosphorus pools in high-elevation lakes of the Sierra Nevada (California): Implications for internal phosphorus loading. *Aquatic Sci.* 76: 511–25.

Homyak, P. M., J. O. Sickman, A. E. Miller, J. M. Melack, T. Meixner, and J. P. Schimel. 2014. N saturation in a chaparral watershed: limitations of N saturation indicators. *Ecosystems* 17: 1286–1305.

Howat, I., and S. Tulaczyk. 2005. Trends in California's spring snowpack over a half century of climate warming. *Ann. Glaciol.* 40: 151–56.

Huber, N. K., and C. D. Rinehart. 1965. Geologic map of the Devil's Postpile quadrangle, Sierra Nevada, California. USGS quadrangle map: GQ-437. https://doi.org/10.3133/*gq437*.

Hundley, N. 2001. *The Great Thirst: Californians and Water: A History.* Berkeley: University of California Press.

Huntington, G. L., and M. A. Akeson. 1987. Soil resource inventory of Sequoia National Park, central part. University of California, Davis, and National Park Service, U.S. Department of Interior.

Huth, K. A., A. Leydecker, J. O. Sickman, and R. C. Bales. 2004. A two-component hydrograph separation for three high-elevation catchments in the Sierra Nevada, California. *Hydrol. Process.* 189: 1721–33.

Iacobellis, S. F., D. R. Daniel, R. Cayan, J. T. Abatzoglou, and H. A. Mooney. 2016. Climate. In H. Mooney and E. Zavaleta, eds., *Ecosystems of California: A Source Book,* 9–25. Oakland: University of California Press.

Jassby, A. D., J. E. Reuter, R. P. Axler, C. R. Goldman, and S. H. Hackley. 1994. Atmospheric deposition of nitrogen and phosphorus in the annual nutrient load of Lake Tahoe (California-Nevada). *Water Resour. Res.* 30: 2207–16.

Jepsen, S. M., N. P. Molotch, M. W. Williams, K. E. Ritter, and J. O. Sickman. 2012. Interannual variability of snowmelt in the Sierra Nevada and Rocky Mountains, United States: Examples from two alpine watersheds. *Water Resour. Res.* 48: W02529.

Kahl, A., A. Winstral, D. Marks, and J. Dozier. 2014. Using satellite imagery and the distributed Isnobal energy balance model to derive SWE heterogeneity in mountainous basins. In J. W. Pomeroy, P. H. Whitfield and C. Spence, eds., *Putting Prediction in Ungauged Basins into Practice,* 243–53. Ottawa: Canadian Water Resources Association.

Kapnick, S., and A. Hall. 2010. Observed climate-snowpack relationships in California and their implications for the future. *J. Climate* 23: 3446–56.

Kattelmann, R. 1990. Floods in the high Sierra Nevada, California, USA. *IAHS Publ.* 194: 311–17.

———. 1992. Historical floods in the eastern Sierra Nevada. In C. A. Hall, V. Doyle-Jones, and B. Widawski, eds., *The History of Water in the Eastern*

Sierra Nevada, Owens Valley, and White Mountains, 74–86. Berkeley: University of California Press.

———. 1996. Hydrology and water resources. In *Sierra Nevada Ecosystem Project: Final Report to Congress,* 855–920. Davis: University of California, Centers for Water and Wildland Resources.

Kattelmann, R. C., and K. Elder. 1991. Hydrologic characteristics and water balance of an alpine basin in the Sierra Nevada. *Water Resour. Res.* 27: 1553–62.

Kaufman, R. 1964. *Gentle Wilderness, the Sierra Nevada.* San Francisco, CA: Sierra Club.

Kelly, C. A. 1988. Toward improving comparisons of alkalinity generation in lake basins. *Limnol. Oceanogr.* 33: 1635–37.

Kerekes, J. J., A. C. Blouin, and S. T. Beuchamp. 1990. Trophic responses to phosphorus in acidic and non-acidic lakes in Nova Scotia, Canada. *Hydrobiologia* 191: 105–10.

Key, J. R., R. Mahoney, Y. Liu, P. Romanov, M. Tschudi, I. Appel, J. Maslanik, D. Baldwin, X. Wang, and P. Meade. 2013. Snow and ice products from Suomi NPP VIIRS. *J. Geophys. Res.* 118: 12816–30.

Kilpatrick, F. A., and E. D. Cobb. 1985. Measurement of discharge using tracers. In *Techniques of Water-Resources Investigations of the United States Geologic Survey,* chap. A16. Reston, VA: U.S. Geological Survey.

Knapp, R. A. 1996. Nonnative trout in natural lakes of the Sierra Nevada: An analysis of their distribution and impacts on native aquatic biota. In *Sierra Nevada Ecosystem Project: Final Report to Congress,* vol. 3, 363–407. Davis: Centers for Water and Wildland Resources, University of California.

Knapp, R. A., D. M. Boiano, and V. T. Vredenburg. 2007. Removal of nonnative fish results in population expansion of a declining amphibian (mountain yellow-legged frog, *Rana muscosa*). *Biol. Conserv.* 135: 11–20.

Knapp, R. A., P. S. Corn, and D. E. Schindler. 2001. The introduction of nonnative fish into wilderness lakes: Good intentions, conflicting mandates, and unintended consequences. *Ecosystems* 4: 275–78.

Knapp, R. A., C. P. Hawkins, J. Ladau, and J. G. McClory. 2005. Fauna of Yosemite National Park lakes has low resistance but high resilience to fish introductions. *Ecol. Appl.* 15: 835–47.

Knapp, R. A., and K. R. Matthews. 2000. Non-native fish introductions and the decline of the mountain yellow-legged frog from within protected areas. *Conserv. Biol.* 14: 428–38.

Knapp, R. A., K. R. Matthews, and O. Sarnelle. 2001. Resistance and resilience of alpine lake fauna to fish introductions. *Ecol. Monogr.* 71: 410–21.

Knapp, R. A., and O. Sarnelle. 2008. Recovery after local extinction: Factors affecting re-establishment of alpine lake zooplankton. *Ecol. Appl.* 18: 1850–59.

Kopacek, J., J. Borovec, J. Hejzlar, K. U. Ulrich, S. A. Norton, and A. Amirbahman. 2005. Aluminum control of phosphorus sorption by lake sediments. *Environ. Sci. Tech.* 39: 8784–89.

Kostadinov, T. S., R. Schumer, M. Hausner, K. J. Bormann, R. Gaffney, K. McGwire, T. H. Painter, S. Tyler, and A. A. Harpold. 2019. Watershed-scale mapping of fractional snow cover under conifer forest canopy using lidar. *Remote Sens. Environ.* 222: 34–49.

Kratz, K., S. D. Cooper, and J. M. Melack. 1994. Effects of single and repeated experimental acid pulses on invertebrates in a high altitude Sierra Nevada stream. *Freshwater Biol.* 32: 161–83.

Landers, D. H., J. M. Eilers, D. F. Brakke, W. S. Overton, P. E. Kellar, M. E. Silverstein, R. D. Schonbrod, R. E. Crowe, R. A. Linthurst, J. M. Omernik, S. A. Teague, and E. P. Meier. 1987. *Characteristics of Lakes in the Western United States,* vol. 1, *Population Descriptions and Physico-Chemical Relationships.* EPA/600/3–86/054a. Washington, DC: U.S. Environmental Protection Agency.

Lettenmaier, D. P., D. Alsdorf, J. Dozier, G. J. Huffman, M. Pan, and E. F. Wood. 2015. Inroads of remote sensing into hydrologic science during the WRR era. *Water Resour. Res.* 51: 7309–42.

Levy, R. 1978. Eastern Miwok. In R. F. Heizer, ed., *Handbook of North American Indians,* vol. 8, *California,* 398–413. Washington, DC: Smithsonian Institution.

Leydecker, A., and J. M. Melack. 1999. Evaporation from snow in the central Sierra Nevada of California. *Nordic Hydrol.* 302: 81–108.

———. 2000. Estimating evaporation in seasonally snow-covered catchments in the Sierra Nevada. *J. Hydrology* 236 121–38.

Leydecker, A., J. O. Sickman, and J. M. Melack. 1999. Episodic lake acidification in the Sierra Nevada, California. *Water Resour. Res.* 359: 2793–2804.

———. 2001. Spatial scaling of hydrological and biogeochemical aspects of high-altitude catchments in the Sierra Nevada, California, U.S.A. *Arctic, Antarctic Alpine Res.* 33: 391–96.

Likens, G. E., F. H. Bormann, R. S. Pierce, J. S. Eaton, and N. M. Johnson. 1977. *Biogeochemistry of a Forested Ecosystem.* New York: Springer.

Liljeblad, S., and C. S. Fowler. 1986. Owens +Valley Paiute. In W. L. D'Azevedo, ed., *Handbook of North American Indians,* vol. 11, *Great Basin,* 412–34. Washington, DC: Smithsonian Institution.

Lockwood, J. P., and P. A. Lydon. 1975. Geologic map of the Mount Abbot quadrangle, central Sierra Nevada, California. USGS quadrangle map: GQ-1155. Denver, CO: USGS Information Services.

Luce, C. H., J. T. Abatzoglou, and Z. A. Holden. 2013. The missing mountain water: Slower Westerlies decrease orographic enhancement in the Pacific Northwest USA. *Science* 342: 1360–64.

Lund, L. J., and A. D. Brown. 1989. Integrated soil processes studies at Emerald Lake Watershed. Final Report, Contract A5–204–32. California Air Resources Board.

Lundquist, J. D., J. R. Minder, P. J. Neiman, and E. Sukovich. 2010. Relationships between barrier jet heights, orographic precipitation gradients, and streamflow in the northern Sierra Nevada. *J. Hydrometeorology* 11: 1141–56.

MacIntyre, S., and J.M. Melack. 2009. Mixing dynamics in lakes across climatic zones. In *Encyclopedia of Ecology of Inland Water,* vol. 2, 603–12. New York: Elsevier.

Magnuson, J.J., D.M. Robertson, B.J., Benson, R.H. Wynne, D.M. Livingstone, T. Arai, R.A. Assel, R.G. Barry, V. Card, E. Kuusisto, N.G. Granin, T.D. Prowse, K.M. Stewart, and V.S. Vuglinski. 2000. Historical trends in lake and river ice cover in the Northern Hemisphere. *Science* 289: 1743–47; errata 2001 *Science* 291: 254.

Majors, J., and D.W. Taylor. 1977. Alpine. In M.G. Barbour and J. Majors, eds., *Terrestrial Vegetation of California,* 601–75. New York: John Wiley and Sons.

Margulis, S.A., G. Cortés, M. Girotto, and M. Durand. 2016a. A Landsat-era Sierra Nevada snow reanalysis (1985–2015). *J. Hydrometeorology* 17: 1203–21.

Margulis, S.A., G. Cortés, M. Girotto, L.S. Huning, D. Li, and M. Durand. 2016b. Characterizing the extreme 2015 snowpack deficit in the Sierra Nevada (USA) and the implications for drought recovery. *Geophys. Res. Lett.* 43: 6341–49.

Margulis, S.A., Y. Fang, D. Li, D.P. Lettenmaier, and K. Andreadis. 2019. The utility of infrequent snow depth images for deriving continuous space-time estimates of seasonal snow water equivalent. *Geophys. Res. Lett.* 46: 5331–40.

Marks, D., J. Domingo, D. Susong, T. Link and D. Garen. 1999. A spatially distributed energy balance snowmelt model for application in mountain basins. *Hydrol. Process.* 13: 1935–59.

Marks, D., and J. Dozier. 1979. A clear-sky longwave radiation model for remote alpine areas. *Theor. Appl. Climatol.* 27: 159–87.

———. 1992. Climate and energy exchange at the snow surface in the alpine region of the Sierra Nevada, 2, Snow cover energy balance. *Water Resour. Res.* 28: 3043–54.

Marks, D., A. Winstral, M. Reba, J. Pomeroy, and M. Kumar. 2013. An evaluation of methods for determining during-storm precipitation phase and the rain/snow transition elevation at the surface in a mountain basin. *Adv. Water Resour.* 55: 98–110.

Marks, D., A. Winstral, G. Flerchinger, M. Reba, J. Pomeroy, T. Link, and K. Elder. 2008. Comparing simulated and measured sensible and latent heat fluxes over snow under a pine canopy to improve an energy balance snowmelt model. *J. Hydrometeorology* 9: 1506–22.

Marshall, A.M., J.T. Abatzoglou, T.E. Link, and C.J. Tennant. 2019. Projected changes in interannual variability of peak snowpack amount and timing in the western United States. *Geophys. Res. Lett.* 46: 8882–92.

Martinec, J., and A. Rango. 1981. Areal distribution of snow water equivalent evaluated by snow cover monitoring. *Water Resour. Res.* 17: 1480–88.

Matthews, K.R., R.A. Knapp, and K.L. Pope. 2002. Garter snake distributions in high-elevation aquatic ecosystems: Is there a link with declining amphibian populations and nonnative trout introductions? *J. Herpetology* 36: 16–22.

McDonald, C.P., J.A. Rover, E.G. Stets, and R.G. Striegl. 2012. The regional abundance and size distribution of lakes and reservoirs in the United States and implications for estimates of global lake extent. *Limnol. Oceanogr.* 57: 597–606.

McKnight, D. M., E. W. Boyer, P. K. Westerhoff, P. T. Doran, T. Kulbe, and D. T. Andersen. 2001. Spectrofluorometric characterization of dissolved organic matter for indication of precursor organic material and aromaticity. *Limnol. Oceanogr.* 46: 38–48.

Meixner, T., and R. C. Bales. 2003. Hydrochemical modeling of coupled C and N cycling in high-elevation catchments: Importance of snow cover. *Biogeochemistry* 62: 289–308.

Meixner, T., C. K. Gutmann, R. C. Bales, A. Leydecker. J. Sickman, J. M. Melack, and J. McConnell. 2004. Multidecadal hydrochemical response of a Sierra Nevada watershed: Sensitivity to weathering rate and changes in deposition. *J. Hydrology* 285: 272–85.

Melack, J. M., S. D. Cooper, and R. W. Holmes. 1987. Chemical and biological survey of lakes and streams located in the Emerald Lake watershed: Sequoia National Park. Final Report, Contract A3–096–32. California Air Resources Board.

Melack, J. M., S. D. Cooper, and T. M. Jenkins Jr. 1989. Biological and chemical characteristics of Emerald Lake and the streams in its watershed, and responses of the lakes and streams to acidic deposition. Final Report, Contract A6–184–37. California Air Resources Board.

Melack, J. M., J. Dozier, C. R. Goldman, D. Greenland, A. Milner, and R. J. Naiman. 1997. Effects of climate change on inland waters of the Pacific coastal mountains and western Great Basin of North America. *Hydrol. Process.* 11: 971–92.

Melack, J. M., S. K. Hamilton, and J. O. Sickman. 1993. Interannual solute variations in a high elevation lake of the Sierra Nevada, California. *Verh. Internat. Verein. Limnol.* 25: 374–77.

Melack, J. M., R. Jellison, S. MacIntyre, and J. T. Hollibaugh. 2017. Mono Lake. In R. Gulati, A. Degermendzhy, and E. Zadereev, eds., *Ecology of Meromictic Lakes,* 325–51. Ecological Studies. New York: Springer.

Melack, J. M. and G. Schladow. 2016. Lakes. In H. Mooney and E. Zavaleta, eds., *Ecosystems of California: A Source Book,* 693–712. Oakland: University of California Press.

Melack, J. M., and J. Sickman. 1995. Snowmelt induced changes in the chemistry of seven high-elevation streams, Sierra Nevada, California. *IAHS Publ.* 228: 221–34.

Melack, J. M., J. O. Sickman, A. Leydecker, and D. Marrett. 1998. Comparative analyses of high-altitude lakes and catchments in the Sierra Nevada: Susceptibility to acidification. Final Report, Contract A032–188. California Air Resources Board.

Melack, J. M., J. O. Sickman, F. Setaro, and D. Dawson. 1994. Monitoring of wet deposition in alpine areas in the Sierra Nevada. Final Report, Contract A932–081. California Air Resources Board.

Melack, J. M., J. O. Sickman, F. Setaro, and D. Dawson. 1997. Monitoring of wet deposition in alpine areas in the Sierra Nevada. Final Report, Contract A932–081. California Air Resources Board.

Melack, J. M., J. O. Sickman, F. V. Setaro, and D. Engle. 1993. Long-term studies of lakes and watersheds in the Sierra Nevada: Patterns and processes of surface-water acidification. Final Report, Contract A932–060. California Air Resources Board.

Melack, J.M., and J.L. Stoddard. 1991. Sierra Nevada. In D.F. Charles, ed., *Acidic Deposition and Aquatic Ecosystems: Regional Case Studies,* 503–30. New York: Springer.

Melack, J.M., J.L. Stoddard, and C.A. Ochs. 1985. Major ion chemistry and sensitivity to acid precipitation of Sierra Nevada lakes. *Water Resour. Res.* 21: 27–32.

Meromy, L., N.P. Molotch, T.E. Link, S.R. Fassnacht, and R. Rice. 2013. Subgrid variability of snow water equivalent at operational snow stations in the western USA. *Hydrol. Process.* 27: 2383–2400.

Meyers, T.P., P. Finkelstein, J. Clarke, T.G. Ellestad, and P.F. Sims. 1998. A multi-layer model for inferring dry deposition using standard meteorological measurements. *J. Geophys. Res.: Atmosph.* 103: 22645–61.

Miller, A.E., J.P. Schimel, J.O. Sickman, T. Meixner, A.P. Doyle, and J.M. Melack. 2007. Mineralization responses at near-zero temperatures in three alpine soils. *Biogeochemistry* 84: 233–45.

Miller, A.E., J.P. Schimel, J.O. Sickman, K. Skeen, T. Meixner, and J.M. Melack. 2009. Seasonal variation in nitrogen uptake and turnover in two high-elevation soils: Mineralization responses are site-dependent. *Biogeochemistry* 93: 253–70.

Miller, M.P., and D.M. McKnight. 2015. Limnology of the Green Lakes Valley: Phytoplankton ecology and dissolved organic matter biogeochemistry at a long-term ecological research site. *Plant Ecol. Diversity* 8: 689–702.

Molot, L.A., P.J. Dillon, and B.D. LaZerte. 1989. Factors affecting alkalinity concentrations of streamwater during snowmelt in central Ontario. *Can. J. Fish Aquat. Sci.* 46: 1658–66.

Molotch, N.P., and R.C. Bales. 2005. Scaling snow observations from the point to the grid element: Implications for observation network design. *Water Resour. Res.* 41: W11421.

Molotch, N.P., M.T. Colee, R.C. Bales, and J. Dozier. 2005. Estimating the spatial distribution of snow water equivalent in an alpine basin using binary regression tree models: The impact of digital elevation data and independent variable selection. *Hydrol. Process.* 19: 1459–79.

Molotch, N.P., and S.A. Margulis. 2008. Estimating the distribution of snow water equivalent using remotely sensed snow cover data and a spatially distributed snowmelt model: A multi-resolution, multi-sensor comparison. *Adv. Water Resour.* 31: 1503–14.

Molotch, N.P., and L. Meromy. 2014. Physiographic and climatic controls on snow cover persistence in the Sierra Nevada Mountains. *Hydrol. Process.* 28: 4573–86.

Montoya, E.L., J. Dozier, and W. Meiring. 2014. Biases of April 1 snow water equivalent records in the Sierra Nevada and their associations with large-scale climate indices. *Geophys. Res. Lett.* 41: 5912–18.

Moore, J.G., and G.S. Mack. 2008. Map showing limits of Tahoe glaciation in Sequoia and Kings Canyon National Parks, California. U.S. Geological Survey Scientific Investigations Map 2945. https://pubs.usgs.gov/sim/2945/sim2945.pdf.

Moore, J. G., and B. C. Moring. 2013. Rangewide glaciation in the Sierra Nevada, California. *Geosphere* 9: 1804–18.

Morris, D. P., and W. M. Lewis. 1988. Phytoplankton nutrient limitation in Colorado mountain lakes. *Freshwater Biol.* 20: 315–27.

Moser, K. A., J. S. Baron, J. Brahney, I. A. Oleksy, J. E. Saros, E. J. Hundey, S. A. Sadro, J. Kopáček, R. Sommaruga, M. J. Kainz, A. L. Strecker, S. Chandra, D. M. Walters, D. L. Preston, N. Michelutti, F. Lepori, S. A. Spaulding, K. R. Christianson, J. M. Melack, and J. P. Smol. 2019. Mountain lakes: Eyes on global environmental change. *Global and Planetary Change* 178: 77–95.

Mote, P. W., A. F. Hamlet, M. P. Clark, and D. P. Lettenmaier. 2005. Declining mountain snowpack in western North America. *Bull. Amer. Meteorol. Soc.* 86: 39–49.

Moyle, P. B., R. M. Yoshiyama, and R. A. Knapp. 1996. Status of fish and fisheries. In *Sierra Nevada Ecosystem Project: Final Report to Congress,* vol. 2, 953–73. Davis: Centers for Water and Wildland Resources, University of California.

Mulch, A., S. A. Graham, and C. P. Chamberlain. 2006. Hydrogen isotopes in Eocene river gravels and paleoelevation of the Sierra Nevada. *Science* 313: 87–89.

Mulch, A., A. M. Sarna-Wojcicki, M. E. Perkins, and C. P. Chamberlain. 2008. A Miocene to Pleistocene climate and elevation record of the Sierra Nevada (California). *Proc. Nat. Acad. Sci.* 105: 6819–24.

Musselman, K. N., M. P. Clark, C. Liu, K. Ikeda, and R. Rasmussen. 2017. Slower snowmelt in a warmer world. *Nature Clim. Change* 7: 214–19.

Nanus, L., D. W. Clow, J. E. Saros, V. C. Stephens, and D. H. Campbell. 2012. Mapping critical loads of nitrogen deposition for aquatic ecosystems in the Rocky Mountains, USA. *Environ. Poll.* 166: 125–35.

National Research Council. 1991. *Opportunities in the Hydrologic Sciences.* Washington, DC: National Academies Press.

———. 2013. *Landsat and Beyond: Sustaining and Enhancing the Nation's Land Imaging Program.* Washington, DC: National Academies Press.

Nelson, C. E. 2009. Phenology of high-elevation pelagic bacteria: The roles of meteorologic variability, catchment inputs, and thermal stratification in structuring communities. *ISME J.* 3: 13–30.

Nelson, C. E., and C. A. Carlson 2011. Differential response of high-elevation planktonic bacterial community structure and metabolism to experimental nutrient enrichment. *PLoS ONE* 6: e18320. doi:10.1371/journal.pone.0018320.

Nelson, C. E., S. Sadro, and J. M. Melack. 2009. Contrasting the influence of stream inputs and landscape position on bacterioplankton community structure and dissolved organic matter composition in high-elevation lake chains. *Limnol. Oceanogr.* 54: 1292–1305.

O'Reilly, C., S. Sharma, D. K. Gray, S. E. Hampton, J. S. Read, R. J. Rowley, P. Schneider, J. D. Lenters, P. B. McIntyre, B. M. Kraemer, G. A. Weyhenmeyer, D. Straile, B. Dong, R. Adrian, M. G. Allan, O. Anneville, L. Arvola, J. Austin, J. L. Bailey, J. S. Baron, J. D. Brookes, E. de Eyto, M. T. Dokulil, D. P. Hamilton, K. Havens, A. L. Hetherington, S. N. Higgins, S. Hook, L. R. Izmest'eva, K. D. Joehnk, K. Kangur, P. Kasprzak, M. Kumagai, E. Kuusisto, G. Leshkevich, D. M.

Livingstone, S. MacIntyre, L. May, J. M. Melack, D. C. Mueller-Navarra, M. Naumenko, P. Noges, T. Noges, R. P. North, P.-D. Plisnier, A. Rigosi, A. Rimmer, M. Rogora, L. G. Rudstam, J. A. Rusak, N. Salmaso, N. R. Samal, D. E. Schindler, S. G. Schladow, M. Schmid, S. R. Schmidt, E. Silow, M. E. Soylu, K. Teubner, P. Verburg, A. Voutilainen, A. Watkinson, C. E. Williamson, and G. Zhang. 2015. Rapid and highly variable warming of lake surface waters around the globe. *Geophys. Res. Lett.* 42: 10773–81.

Overturf, T. D. 1991. Hydrogeology of the upper Mammoth Creek drainage basin, Mono County, California. Department of Geological Sciences, California State University, Fullerton.

Painter, T. H., D. F. Berisford, J. W. Boardman, K. J. Bormann, J. S. Deems, F. Gehrke, A. Hedrick, M. Joyce, R. Laidlaw, D. Marks, C. Mattmann, B. McGurk, P. Ramirez, M. Richardson, S. M. Skiles, F. C. Seidel, and A. Winstral. 2016. The Airborne Snow Observatory: Fusion of scanning lidar, imaging spectrometer, and physically-based modeling for mapping snow water equivalent and snow albedo. *Remote Sens. Environ.* 184: 139–52.

Painter, T. H., K. Rittger, C. McKenzie, P. Slaughter, R. E. Davis, and J. Dozier. 2009. Retrieval of subpixel snow-covered area, grain size, and albedo from MODIS. *Remote Sens. Environ.* 113: 868–79.

Pandey, G. R., D. R. Cayan, and K. P. Georgakakos. 1999. Precipitation structure in the Sierra Nevada of California during winter. *J. Geophys. Res.* 104: 12019–30.

Park, S., M. T. Brett, A. Müller-Solger, and C. R. Goldman. 2004. Climatic forcing and primary productivity in a subalpine lake: Interannual variability as a natural experiment. *Limnol. Oceanogr.* 49: 614–19.

Parker, B. R., R. D. Vinebrooke, and D. W. Schindler. 2008. Recent climate extremes alter alpine lake ecosystems. *Proc. Nat. Acad. Sci.* 105: 12927–31.

Patten, D. T., F. P. Conte, W. E. Cooper, J. Dracup, S. Dreiss, K. Harper, G. L. Hunt, P. Kilham, H. E. Klieforth, J. M. Melack, and S. A. Temple. 1987. *The Mono Basin Ecosystem: Effects of Changing Lake Level.* Washington, DC: National Academies Press.

Perrot, D., N. P. Molotch, M. W. Williams, S. M. Jepsen, and J. O. Sickman. 2014. Relationships between stream nitrate concentration and spatially distributed snowmelt in high-elevation catchments of the western U.S. *Water Resour. Res.* 50: 8694–8713.

Peterson, D., M. Arbaugh, V. Wakefield, and P. Miller. 1987. Evidence of growth reduction in ozone-injured Jeffrey pines (*Pinus jeffreyi* Grev. and Balf.) in Sequoia and Kings Canyon National Parks. *J. Air Poll. Control Assoc.* 37: 906–12.

Pierce, D. W., and D. R. Cayan. 2013. The uneven response of different snow measures to human-induced climate warming. *J. Climate* 26: 4148–67.

Pierce, D. W., D. R. Cayan, and B. L. Thrasher. 2014. Statistical downscaling using Localized Constructed Analogs (LOCA). *J. Hydrometeorology* 15: 2558–85.

Piovia-Scott, J., S. Sadro, R. A. Knapp, J. Sickman, K. L. Pope, and S. Chandra. 2016. Variation in reciprocal subsidies between lakes and land: Perspectives from the mountains of California. *Can. J. Fish. Aquat. Sci.* 73: 1691–1701.

Pister, E. P. 2001. Wilderness fish stocking: History and perspective. *Ecosystems* 4: 279–86.

Porinchu, D. F., A. P. Potitio, G. M. MacDonald, and A. M. Bloom. 2007. Subfossil chironomids as indicators of recent climate change in Sierra Nevada, California, lakes. *Arctic Antarctic Alpine Res.* 39: 286–96.

Preston, D. L., N. Caine, D. M. McKnight, M. W. Williams, K. Hell, M. P. Miller, S. J. Hart, and P. T. J. Johnson. 2016. Climate regulates alpine lake ice cover phenology and aquatic ecosystem structure. *Geophys. Res. Lett.* 43: 5353–60.

Psenner, R. 1988. Alkalinity generation in a soft-water lake: Watershed and in-lake processes. *Limnol. Oceanogr.* 33: 1463–75.

Pupacko, A. 1993. Variations in northern Sierra Nevada streamflow: Implications of climate change. *Water Resour. Bull.* 29: 283–90.

Rachowicz, L. J., R. A. Knapp, J. A. T. Morgan, M. J. Stice, V. T. Vredenburg, J. M. Parker, and C. J. Briggs. 2006. Emerging infectious disease as a proximate cause of amphibian mass mortality. *Ecology* 87: 1671–83.

Raleigh, M. S., and J. D. Lundquist. 2012. Comparing and combining SWE estimates from the SNOW-17 model using PRISM and SWE reconstruction. *Water Resour. Res.* 48: W01506.

Rauscher, S. A., J. S. Pal, N. S. Diffenbaugh, and M. M. Benedetti. 2008. Future changes in snowmelt-driven runoff timing over the western US. *Geophys. Res. Lett.* 35: L16703.

Reba, M. L., T. E. Link, D. Marks, and J. Pomeroy. 2009. An assessment of corrections for eddy covariance measured turbulent fluxes over snow in mountain environments. *Water Resour. Res.* 45: W00D38.

Redmond, K. T., and R. W. Koch. 1991. Surface climate and streamflow variability in the western United States and their relationship to large-scale circulation indices. *Water Resour. Res.* 27: 2381–99.

Reimers, N., J. A. Maciolek, and E. P. Pister. 1955. Limnological study of the lakes in Convict Creek Basin, Mono County, California. U.S. Fish and Wildlife Service. *Fishery Bull.* 56: 437–503.

Rice, R., and R. C. Bales. 2010. Embedded sensor network design for snowcover measurements around snow-pillow and snow-course sites in the Sierra Nevada of California. *Water Resour. Res.* 46: W03537.

Rice, R., R. C. Bales, T. H. Painter, and J. Dozier. 2011. Snow water equivalent along elevation gradients in the Merced and Tuolumne River basins of the Sierra Nevada. *Water Resour. Res.* 47: W08515.

Rittger, K., A. Kahl, and J. Dozier. 2011. Topographic distribution of snow water equivalent in the Sierra Nevada. *Proc. Western Snow Conf.* 79: 37–46.

Rittger, K., T. H. Painter, and J. Dozier. 2013. Assessment of methods for mapping snow cover from MODIS. *Adv. Water Resour.* 51: 367–80.

Rose, N. L., E. Shilland, T. Berg, K. Hanselmann, R. Harriman, K. Koinig, U. Nickus, B. S. Trad, E. Stuchlik, H. Thies, and M. Ventura. 2001. Relationships between acid ions and carbonaceous fly-ash particles in deposition at European mountain lakes. *Water Air Soil Poll.* 130: 1703–8.

Rosenfeld, D., W. L. Woddley, D. Axisa, E. Freud, J. G. Hudson, and A. Givati. 2008. Aircraft measurements of the impacts of pollution aerosols on clouds and precipitation over the Sierra Nevada. *J. Geophys. Res.* 113: D215203. doi:10.1029/2007/JD009544.

Rosenthal, J., ed. 2011. *A New Frame of Reference: Prehistoric Cultural Chronology and Ecology in North-Central Sierra Nevada.* Publication 16. Davis: Center for Archaeological Research, University of California.

Rosenthal, W., J. Saleta, and J. Dozier. 2007. Scanning electron microscopy of impurity structures in snow. *Cold Regions Sci. Tech.* 47: 80–89.

Rundel, P. W., M. Neuman, and P. Rabenold. 2009. Plant communities and floristic diversity of the Emerald Lake basin, Sequoia National Park, California. *Madroño* 56: 184–98.

Rundel, P. W., D. J. Parsons, and D. T. Gordon. 1977. Montane and subalpine vegetation of the Sierra Nevada and Cascade ranges. In M. G. Barbour and J. Majors, eds., *Terrestrial Vegetation of California,* 559–99. New York: John Wiley and Sons.

Rundel, P. W., T. V. St. John, and W. L. Berry. 1988. Vegetation process studies. Final Report, Contract A4–121–32. California Air Resources Board.

Sadro, S., and J. M. Melack. 2012. The effect of an extreme rain event on the biogeochemistry and ecosystem metabolism of an oligotrophic high-elevation lake (Emerald Lake, Sierra Nevada, California). *Alpine, Arctic Antarctic Res.* 44: 222–31.

Sadro, S., J. M. Melack, and S. MacIntyre. 2011a. Depth-integrated estimates of ecosystem metabolism in a high-elevation lake (Emerald Lake, Sierra Nevada, California). *Limnol. Oceanogr.* 56: 1764–80.

———. 2011b. Spatial and temporal variability in ecosystem metabolism: free-water and incubation chamber measurements from benthic and pelagic habitats in a high-elevation lake (Emerald Lake, Sierra Nevada, California). *Ecosystems* doi: 10.1007/s10021-011-9471-5.

Sadro, S., J. M. Melack, J. O. Sickman, and K. Skeen. 2019. Response of mountain lakes to climate change: Predicting warming from loss of snow in the Sierra Nevada of California. *Limnol. Oceanogr. Lett.* 4: 9–17.

Sadro, S., C. E. Nelson, and J. M. Melack. 2011. Linking diel patterns in community respiration to bacterioplankton in an oligotrophic high-elevation Sierra Nevada (California, USA) lake. *Limnol. Oceanogr.* 56: 540–50.

———. 2012. The influence of landscape position and catchment characteristics on aquatic biogeochemistry in high-elevation lake chains. *Ecosystems.* doi: 10.1007/s10021-011-9515-x.

Sadro, S., J. O. Sickman, J. M. Melack, and K. Skeen. 2018. Effects of climate variability on snowmelt and implications for organic matter in a high elevation lake. *Water Resour. Res.* doi:10.1029/2017WR022163.

Sarnelle, O., and R. A. Knapp. 2004. Zooplankton recovery after fish removal: Limitations of the egg bank. *Limnol. Oceanogr.* 49: 1382–92.

———. 2005. Nutrient recycling by fish versus zooplankton grazing as drivers of the trophic cascade in alpine lakes. *Limnol. Oceanogr.* 50: 2032–42.

Schindler, D. E., R. A. Knapp, and P. R. Leavitt. 2001. Alteration of nutrient cycles and algal production resulting from fish introductions into mountain lakes. *Ecosystems* 4: 308–21.

Schindler, D. W. 1988. Confusion over the origin of alkalinity in lakes. *Limnol. Oceanogr.* 33: 1637–40.

Schindler, D. W., M. A. Turner, P. Stainton, and G. A. Linsey. 1986. Natural sources of acid neutralizing capacity in low alkalinity lakes of the Precambrian Shield. *Science* 232: 844–47.

Seidel, F. C., K. Rittger, S. M. Skiles, N. P. Molotch, and T. H. Painter. 2016. Case study of spatial and temporal variability of snow cover, grain size, albedo and radiative forcing in the Sierra Nevada and Rocky Mountain snowpack derived from imaging spectroscopy. *Cryosphere* 10: 1229–44.

Selters, A. 1980. *High Sierra Hiking Guide No. 20: Triple Divide Peak*. Berkeley, CA: Wilderness Press.

Sharp, R. P. 1976. *Geology: Field Guide to Southern California*. Dubuque, IA: Kendal/Hunt.

Shaw, G. D., R. Cisneros, D. Schweizer, J. O. Sickman, and M. E. Fenn. 2013. Critical loads of acid deposition for wilderness lakes in the Sierra Nevada (California) estimated by the steady state water chemistry model. *Water Air Soil Poll.* 224: 1804. doi10.1007/s11270–013–1804-x.

Shaw, J. R. 1997. Modeling of silicate mineral weathering reactions in an alpine basin of the southern Sierra Nevada, California. MSc thesis, University of Arizona.

Sickman, J. O. 2001. Comparative analyses of nitrogen biogeochemistry in high elevation ecosystems. PhD dissertation, University of California, Santa Barbara.

Sickman, J. O., D. M. Bennett, D. M. Lucero, T. J. Whitmore, and W. F. Kenney. 2013. Diatom-inference models for acid neutralizing capacity and nitrate based on 41 calibration lakes in the Sierra Nevada, California, USA. *J. Paleolimnol.* 50: 159–74.

Sickman, J. O., A. Leydecker, C. Y. Chang, C. Kendall, J. M. Melack, D. M. Lucero, and J. Schimel. 2003. Mechanisms underlying export of N from high-elevation catchments during seasonal transitions. *Biogeochemistry* 64: 1–24.

Sickman, J. O., A. Leydecker, and J. M. Melack. 2001. Nitrogen mass balances and abiotic controls on N retention and yield in high-elevation catchments of the Sierra Nevada, California, United States. *Water Resour. Res.* 37: 1445–61.

Sickman, J. O., and J. M. Melack. 1989. Characterization of year-round sensitivity of California's montane lakes to acidic deposition. Final Report, Contract A5–203–32. California Air Resources Board.

———. 1992. Photosynthetic activity of phytoplankton in a high altitude lake (Emerald Lake, Sierra Nevada, California). *Hydrobiologia* 230: 37–48.

———. 1998. Nitrogen and sulfate export from high elevation catchments of the Sierra Nevada, California. *Water Air Soil Poll.* 105: 217–26.

Sickman, J. O., J. M. Melack, and D. Clow. 2003. Evidence for nutrient enrichment of high-elevation lakes in the Sierra Nevada, California. *Limnol. Oceanogr.* 48: 1885–92.

Sickman, J. O., J. M. Melack, and J. L. Stoddard. 2002. Regional analysis of inorganic nitrogen yield and retention in high-elevation ecosystems of the Sierra Nevada and Rocky Mountains. *Biogeochemistry* 57–58: 341–74.

Sierra Nevada Ecosystems Project. 1996. Final Report to Congress. Vol. 1: Assessment summaries and management strategies. University of California, Centers for Water and Wildland Resources, Davis.

Singer, M. B., R. Aalto, L. A. James, N. E. Kilham, J. L. Higson, and S. Ghoshal. 2013. Enduring legacy of a toxic fan via episodic redistribution of California gold mining debris. *Proc. Nat. Acad. Sci.* 110: 18436–41.

Sisson, T. W., and J. G. Moore. 1984. Geology of the Giant Forest–Lodgepole area, Sequoia National Park, CA. USGS Open File Rep. No. 84–254.

Smith, C. R. 1978. Tubatulabal. In R. F. Heizer, ed., *Handbook of North American Indians,* vol. 8, *California,* 437–47. Washington, DC: Smithsonian Institution.

Smith, G. S., ed. 1976. *Mammoth Lake Sierra.* Palo Alto, CA: Genny Smith Books.

Smits, A. P., S. MacIntyre, and S. Sadro. 2020. Snowpack determines relative importance of climate factors driving lake thermal regimes. *Limnol. Oceanogr. Lett.* doi: 9443/10.1002/lol2.10147.

Sokal, R. R., and F. J. Rohlf. 1981. *Biometry.* New York: W. H. Freeman and Co.

Spier, R. F. G. 1978. Monache. In R. F. Heizer, ed., *Handbook of North American Indians,* vol. 8, *California,* 426–36. Washington, DC: Smithsonian Institution.

Stauffer, R. E. 1990. Granite weathering and the sensitivity of alpine lakes to acid deposition. *Limnol. Oceanogr.* 35: 1112–34.

Stoddard, J. L. 1987a. Alkalinity dynamics in an unacidified alpine lake, Sierra Nevada, California. *Limnol. Oceanogr.* 32: 825–39.

———. 1987b. Microcrustacean communities of high elevation lakes in the Sierra Nevada, California. *J. Plankton Res.* 9: 631–50.

———. 1987c. Micronutrient and phosphorus limitation of phytoplankton abundance in Gem Lake, Sierra Nevada, California. *Hydrobiologia* 154: 103–11.

———. 1988. Are Sierran lakes different? A reply to the comments of Kelly and Schindler. *Limnol. Oceanogr.* 33: 1641–46.

———. 1994. Long-term changes in watershed retention of nitrogen. Its causes and aquatic consequences. In L. A. Baker, ed., *Environmental Chemistry of Lakes and Reservoirs,* 223–84. Washington, DC: American Chemical Society.

———. 1995. Episodic acidification during snowmelt of high elevation lakes in the Sierra Nevada mountains of California. *Water Air Soil Poll.* 85: 353–58.

Stoddard, J. L., and J. O. Sickman. 2002. Episodic acidification of lakes in the Sierra Nevada. Final Report, Contract A132–048. California Air Resources Board.

Stoddard, J. L., J. Van Sickle, A. T. Herlihy, J. Brahney, S. Paulsen, D. V. Peck, R. Mitchell, and A. I. Pollard. 2016. Continental-scale increase in lake and stream phosphorus: Are oligotrophic systems disappearing in the United States? *Environ. Sci. Tech.* 50: 3409–15.

Stohlgren, T. J., J. M. Melack, A. L. Esperanza, and D. J. Parsons. 1991. Atmospheric deposition and solute export in Giant Sequoia–mixed conifer watersheds in the Sierra Nevada, California. *Biogeochemistry* 12: 207–30.

Stohlgren, T. J., and D. J. Parsons. 1987. Variation of wet deposition chemistry in Sequoia National Park, California. *Atmosph. Environ.* 21: 1369–74.

Storer, T. I., R. L. Usinger, and D. Lukas. 2004. *Sierra Nevada Natural History.* Berkeley: University of California Press.

Suchet, P. A., and J. L. Probst. 1993. Modelling of atmospheric CO_2 consumption by chemical weathering of rocks: Application to the Garonne, Congo and Amazon basins. *Chem. Geol.* 107: 205–10.

Urbanski, S. P., W. M. Hao, and S. Baker. 2009. Chemical composition of wildland fire emissions. In A. Bytnerowicz, M. Arbaugh, C. Andersen and A. Riebau, eds., *Wildland Fires and Air Pollution,* 79–107. Developments in Environmental Science 8. Amsterdam: Elsevier.

Vicars, W. C., and J. O. Sickman. 2011. Mineral dust transport to the Sierra Nevada, California: Loading rates and potential source areas. *J. Geophys. Res.: Biogeosci.* 116: G01018. doi:10.1029/2010JG001394.

Vicars, W. C., J. O. Sickman, and P. J. Ziemann. 2010. Atmospheric phosphorus deposition at a montane site: Size distribution, effects of wildfire, and ecological implications. *Atmosph. Environ.* 44: 2813–21.

Warren, S. G., and R. E. Brandt. 2008. Optical constants of ice from the ultraviolet to the microwave: A revised compilation. *J. Geophys. Res.* 113: D14220.

Whiting, M. C., D. R. Whitehead, R. W. Holmes, and S. A. Norton. 1989. Paleolimnological reconstruction of recent acidity changes in four Sierra Nevada lakes. *J. Paleolimnol.* 2: 285–304.

Whitney, S. 1979. *The Sierra Nevada.* San Francisco, CA: Sierra Club Books.

Williams, M. R., A. Leydecker, A. D. Brown, and J. M. Melack. 2001. Processes regulating the solute concentrations of snowmelt runoff in two subalpine catchments of the Sierra Nevada, California. *Water Resour. Res.* 37: 1993–2008.

Williams, M. R., and J. M. Melack. 1997. Atmospheric deposition, mass balances, and processes regulating streamwater solute concentrations in mixed conifer catchments of the Sierra Nevada, California. *Biogeochemistry* 37: 111–44.

Williams, M. W., R. Bales, A. D. Brown, and J. M. Melack. 1995. Fluxes and transformations of nitrogen in a high-elevation catchment, Sierra Nevada. *Biogeochemistry* 28: 1–31.

Williams, M. W., J. Baron, N. Caine, R. Sommerfeld, and R. Sanford. 1996. Nitrogen saturation in the Colorado Front Range. *Environ. Sci. Tech.* 30: 640–46.

Williams, M. W., A. D. Brown, and J. M. Melack. 1993. Geochemical and hydrological controls on the composition of surface water in a high-elevation basin, Sierra Nevada, California. *Limnol. Oceanogr.* 38: 775–97.

Williams, M. W., and D. W. Clow. 1990. Hydrologic and biologic consequences of an avalanche striking an ice-covered lake. In *Proc. 58th Western Snow Conf.,* 51–60. Brush Prairie, WA: Western Snow Conference.

Williams, M. W., M. Losleben, N. Caine, and D. Greenland. 1996. Changes in climate and hydrochemical responses in a high-elevation catchment, Rocky Mountains. *Limnol. Oceanogr.* 41: 939–46.

Williams, M. W., and J. M. Melack. 1989. Effects of spatial and temporal variation in snowmelt on nitrate and sulfate pulses in an alpine watershed. *Ann. Glaciol.* 13: 285–88.

———. 1991a. Precipitation chemistry and ionic loading to an alpine basin, Sierra Nevada. *Water Resour. Res.* 27: 1563–74.

———. 1991b. Solute chemistry of snowmelt and runoff in an alpine basin, Sierra Nevada. *Water Resour. Res.* 27: 1575–88.

Williams, W. T., M. Brady, and S. C. Willison. 1977. Air pollution damage to the forests of the Sierra Nevada Mountains of California. *J. Air Poll. Control Assoc.* 27: 230–36.

Winstral, A., D. Marks, and R. Gurney. 2013. Simulating wind-affected snow accumulations at catchment to basin scales. *Adv. Water Resour.* 55: 64–79.

Winter, T. C. 1981. Uncertainties in estimating the water balance of lakes. *Water Resour. Bull.* 17: 82–115.

Wiscombe, W. J., and S. G. Warren. 1980. A model for the spectral albedo of snow, I, Pure snow. *J. Atmosph. Sci.* 37: 2712–33.

Wolford, R. A., and R. C. Bales. 1996. Hydrochemical modeling of Emerald Lake watershed, Sierra Nevada, California: Sensitivity of stream chemistry to changes in fluxes and model parameters. *Limnol. Oceanogr.* 41: 947–54.

Wolford, R., R. Bales, and S. Sorooshian. 1996. Development of a hydrochemical model for seasonally snow-covered alpine watersheds: Application to Emerald Lake watershed, Sierra Nevada, California. *Water Resour. Res.* 32: 1061–74.

Wood, E. F., M. Sivapalan, K. Beven, and L. Band. 1988. Effects of spatial variability and scale with implications to hydrological modeling. *J. Hydrology* 102: 29–47.

Yang, D., D. L. Kane, L. D. Hinzman, B. E. Goodison, J. R. Metcalfe, P. Y. T. Louie, G. H. Leavesley, D. G. Emerson, and C. L. Hanson. 2000. An evaluation of the Wyoming gauge system for snowfall measurement. *Water Resour. Res.* 36: 2665–77.

Yang, D., D. Kane, Z. Zhang, D. Legates, and B. Goodison. 2005. Bias corrections of long-term (1973–2004) daily precipitation data over the northern regions. *Geophys. Res. Lett.* 32: L19501.

INDEX

Founded in 1893,
UNIVERSITY OF CALIFORNIA PRESS
publishes bold, progressive books and journals
on topics in the arts, humanities, social sciences,
and natural sciences—with a focus on social
justice issues—that inspire thought and action
among readers worldwide.

The UC PRESS FOUNDATION
raises funds to uphold the press's vital role
as an independent, nonprofit publisher, and
receives philanthropic support from a wide
range of individuals and institutions—and from
committed readers like you. To learn more, visit
ucpress.edu/supportus.